Creative Activities in Mathematics

PROBLEM-BASED MATHS
INVESTIGATIONS FOR LOWER
AND MIDDLE PRIMARY

BOOK 1

BY DEREK HOLTON AND CATH PEARN

Published in 2025 by Amba Press, Melbourne, Australia
www.ambapress.com.au

First published in 2015 by ACER Press, an imprint of
Australian Council for Educational Research Ltd

© 2025 Derek Holton and Cath Pearn

All rights reserved. No part of this book may be reproduced or transmitted in any form or by any means, electronic or mechanical, including photocopying, recording or by any information storage and retrieval system, without prior permission in writing from the publisher.

Cover design: Tess McCabe
Editor: Diane Fowler
Typesetting: Peter Long

ISBN: 9781923569225 (pbk)
ISBN: 9781923569232 (ebk)

A catalogue record for this book is available from the National Library of Australia.

TABLE OF CONTENTS

Introduction: A recipe for success	5
Part 1: Number and Algebra	**9**
Chapter 1: Numbers and words	10
Chapter 2: Cats and dogs	20
Chapter 3: Difference and sum problems	34
Chapter 4: Arks and Tarks	44
Chapter 5: The farmyard problem	56
Chapter 6: The 12 game	65
Part 2: Measurement and Geometry	**77**
Chapter 7: Classroom shapes	78
Chapter 8: Jenny's jelly bean problem	92
Chapter 9: Horrible Hal's humungous hall	99
Part 3: Statistics and Probability	**113**
Chapter 10: The longest name	114
Chapter 11: Penny's Pet Shop	123
Chapter 12: The teddies' race	133

INTRODUCTION: A RECIPE FOR SUCCESS

Who is this book for?

This book aims to provide extended activities that introduce or prepare the way for material from the lower primary section of the Australian Curriculum: Mathematics, or that consolidate material that has already been taught. All of the activities provide valuable opportunities for developing students' abilities within the curriculum's proficiency strands. It is for the use of:

- experienced teachers who work regularly in the area of mathematics
- teachers who may not be as confident in maths as in other areas
- relieving teachers needing additional maths support and class resources.

This is the first of three books in the *Creative Activities in Mathematics* series that cover most of the school years. Although Book 2 is largely for upper primary students, some of the material of that book can be used for lower primary students too. Book 3 is for secondary school students.

Background

The aim of this book is to produce twelve stimulating mathematical activities that can be used with students from Foundation to Year 4 and probably even further. Each of these activities is presented at three levels. In moving from Level 1 to Level 2, and from Level 2 to Level 3, there is an extension of some kind, such as an increase in difficulty or the development of a new concept. We will give the details of these at the start of each chapter. The point of these three levels is to give you, the teacher, the chance to match the activities to your students and to extend their mathematical skills, knowledge and understanding.

This book tries to present lesson ideas that you can use as is, but we hope you will bring your own approach to the lessons and develop them for the students in your class. It's important to consider the activities here as a basis for class work, in much the same way as you might produce a dish from a page of a recipe book. If you think that a Spanish omelette is better with goose eggs than chickens' eggs, use goose eggs. If you feel happier making the omelette with goose eggs, you'll likely produce a better meal as a consequence. So don't be afraid to add or subtract your own variations to the 'recipes' in this book.

It's worth noting that you don't have to make all of each recipe at once; some things are best left to stand for a while. You don't have to do all of what is suggested for any of the levels of each activity in consecutive sessions. Pick out the parts that are most suitable for your students, and come back to the rest at a more appropriate time. Consider what each activity offers and look out for the best time and way to introduce it.

Some things you cook will taste the same no matter what you call them; 'caviar' and 'fish eggs' are much the same thing. It's the food's value and the taste that counts, not how you say it. So, too, it's important for students to understand ideas about shape but whether you say a 'cuboid' or a 'box' isn't as important at this early stage of their development. We generally try to use correct mathematical language, but whether you do or not in your class is up to you.

On the other hand, you have to get the ingredients right. If you tell your guests they're about to have roast pork and you've made it with rabbit, they'll notice the absence of crackling. What is vitally important to the recipes in this book is the involvement of the learner. In all the work here we want the students to think, reason and solve as many of the problems as they can by themselves or with minimal help. Only tell them the answer as a last resort, and give them as much scaffolding as you can before you get to that point.

The activities here are meant to be used in as open a way as possible. We firmly believe that 'the one who does the thinking does the learning'. Give your students the opportunity to think for themselves so that they can understand and learn more deeply. Give them every opportunity to work through the various situations themselves and do so in their own way. And at any point on their journey, don't be afraid to ask them how they got there.

As a result of this philosophy, you'll see phrases such as 'Discuss', 'Let your students', 'Ask' and so on throughout the book. These are a sign that the students' role is to work, think, understand and learn. We expect them to move from discussing the problems as a whole class to working in small groups on the problem details and then reporting their work back to the whole class. In this way they will be introduced to and engaged in the proficiency strands of Understanding, Fluency, Problem Solving and Reasoning, as well as gathering confidence in communicating their ideas to you and their peers.

Your role is to prepare the ingredients and decide on the recipe, but to let the students be involved in the cooking. We hope that you will introduce the students to the ideas of mathematics and gradually give them more and more opportunities to do the actual tasks for themselves.

The activities in this book aim to provide situations where students can learn and practise mathematical skills, but they also aim to provide opportunities for them to think, reason and be creative.

How to use this book

This book has been written for you to use when looking for rich mathematical activities to offer your students in regular classrooms. The material here is founded on the Australian Curriculum: Mathematics. All of the activities develop in some way as they go through their levels. This development may be in terms of the curriculum's content strands, proficiency strands or both.

Your first choice is to look for an activity that fits what you are planning to do or what you have done. An activity may rely on content areas that you have already covered or areas that you're planning to cover.

Your second choice is to identify the levels appropriate for your individual students or groups. Each activity has three levels and they develop in some way from each level to the next. Not all of your students will be able to work past Level 1 or Level 2 of a given activity; it's also possible that you may want to omit Level 3. Therefore there are provisions throughout for differentiation, and to allow all students to work on tasks that are appropriate for their mathematical skills and understandings.

Your third choice is to consider how you might alter some activities to best suit your class. When you look at an activity you might see that the basic idea is sound but want to simplify it (or extend it) in some way. For instance, in the *Numbers and letters* activity (Chapter 2), it may be better for your students to start with a table that uses numbers 1 to 26 rather than the numbers 1 and 2. Or you may feel that larger numbers, decimals or fractions are better for some of your students. You should feel free to make these adjustments as you think fit.

The Australian Curriculum: Mathematics is central to these activities. Each activity includes a table that lists relevant content descriptions. All of the activities here have a strong link to the Problem Solving proficiency strand.

Activity layout

Each activity has the following format.
- Initial problem
- Background information
- Big ideas
- Suggested resources
- Problem aims
- Key concepts
- Possible heuristics/strategies
- Special notes
- Levels and steps

The *Initial problem* is the start of the activity for the students and gives some idea of the topic of the activity. This is followed by *Background information*, which aims to give teachers an overview of the activity and how it develops from the initial problem. It sometimes notes any links with similar problems and related ideas. A table sets out the Australian Curriculum: Mathematics content descriptors for each level of each activity. (Note that only part of some descriptions may be covered by a given activity.) This table also sets out the way the problem develops and makes comments on the mathematics that the levels contain. After the background information we list the *Big ideas* of the activity.

It's probably self-evident what *Suggested resources*, *Problem aims*, *Key concepts* and *Possible heuristics/strategies* are. *Special notes* are included when there is a definition or idea that may not be common at early primary levels and may need to be explained. For instance, in the *Arks and Tarks* problem in Chapter 4, we say what we mean by 'least possible' as it has special importance in the problem.

The activity is then broken up into its three levels, which develop the problem based around questions and answers. Changes of directions or extensions of the problem are indicated by 'steps'. Throughout each problem there are questions that teachers might ask of students in the course of developing the activities.

Each level problem is completed by a section called 'Where to from here?' that provides questions teachers could ask students, focusing on the big ideas involved at that level. This section provides new ideas to follow up and enables you and your class to enjoy yourself thinking up new problems as extensions from the work of that level.

Additional resources are available at the series website: http://www.acer.edu.au/cam. These include references and web links to related material, plus activity sheets for students that provide a framework for their responses. These connect to the problems of that level and can be printed or copied for the students' use. Most of these student materials have a strongly visual focus and should be accessible for all students, even Foundation and Year 1 students with limited literacy skills.

PART 1: NUMBER AND ALGEBRA

Part 1 presents six activities centred on the Number and Algebra strand.

Table 1.1: Number and Algebra activities

Problem	Big ideas
Numbers and words	- Explore and use basic addition, subtraction, multiplication and division facts by 'scoring' words to first give numbers less than 100 and then up to 1000 - Understand the restrictions implied by the scoring of words, or giving letters a value - Use comparative words such as bigger/biggest - Use simple fractions $\frac{1}{5}$ and $\frac{1}{10}$
Cats and dogs	- Construct and use tables and pictograms /graphs to represent data - Identify number patterns - Describe rules using the appropriate language - Use 'multiplication' and 'division' - Relate mathematics to a real-world situation
Difference and sum problems	- Use 2-, 3- and 4-digit addition and subtraction - Look for patterns - Explain patterns (justify)
Arks and Tarks	- Understand the relationship between various aspects of a problem in order to solve it - Record and justify the answers
The farmyard problem	- Understand the relationship between various aspects of a problem in order to solve it - Identify number patterns in order to solve a problem
The 12 game	- Understand the relation between various aspects of a problem in order to solve it - Identify number patterns in order to solve a problem

Some reminders before you use these tasks in your classroom:

1. The questions in the text are ones you can ask your students. You are likely to be able to produce similar, more immediately relevant ones for your particular students as you work on these activities with them.
2. We have given suggested links to the Years in the Australian Curriculum: Mathematics for all the Levels in each activity. Given that there will be a spread of ability in your classes, you should take these as a guide only. Take the opportunity to encourage every student to the edge of their comfort zone.
3. To take all students further, sometimes you can omit some of the later steps of a Level in favour of the early steps in the following Level.

CHAPTER 1: NUMBERS AND WORDS

Initial problem

Ask the students to draw a picture of a cat and write the word CAT underneath. Link the letters of the alphabet to the numbers in whichever table below that you think is best for your students.

A	B	C	D	E	F	G	H	I	J	K	L	M
1	1	1	1	1	1	1	1	1	1	1	1	1
N	O	P	Q	R	S	T	U	V	W	X	Y	Z
1	1	1	1	1	1	1	1	1	1	1	1	1

A	B	C	D	E	F	G	H	I	J	K	L	M
1	2	1	2	1	2	1	2	1	2	1	2	1
N	O	P	Q	R	S	T	U	V	W	X	Y	Z
2	1	2	1	2	1	2	1	2	1	2	1	2

A	B	C	D	E	F	G	H	I	J	K	L	M
1	2	3	1	2	3	1	2	3	1	2	3	1
N	O	P	Q	R	S	T	U	V	W	X	Y	Z
2	3	1	2	3	1	2	3	1	2	3	1	2

A	B	C	D	E	F	G	H	I	J	K	L	M
1	2	3	4	5	6	7	8	9	10	11	12	13
N	O	P	Q	R	S	T	U	V	W	X	Y	Z
14	15	16	17	18	19	20	21	22	23	24	25	26

Ask them to put the right numbers under the letters of CAT. What do the numbers of the word CAT add up to?

Background information

Numbers and words is about finding the score of a given word if numbers are assigned to letters of the alphabet. It also looks at things like 'highest and lowest' and 'biggest and smallest' scores.

We've provided four tables here so that you can choose numbers suitable to the children in your class. You may decide to give different tables to different students, or use one table now and another table later in the year.

It's worth noting that if you use a table where number values repeat, some of the activities will function differently than they would if each letter has a unique number value, particularly at Level 1. For instance, if every letter has a value of 1, then every 3-letter word has the same score (3) and students will need to come up with longer words to create higher scores. Similarly, it will be impossible for students to deduce the 'correct' word from a score if all words of the same length have the same score. Be prepared to adapt the way you use the activities to fit the table of values you use, and to change the way things work if/when you change tables.

At Level 1 we investigate the scores of 3- and 4-letter words using addition. The activity is accessible to all year levels of the curriculum if one of the first three tables is used. Later in the year, as their ability increases, you could use the activity again but with larger numbers. With more able students in your class, you might use higher numbers from the start.

The Level 2 activity uses addition and subtraction with students' names. This activity should be available for Year 2 students and up, but again, using smaller numbers will extend this range downwards.

In Level 3 we restrict ourselves to numbering consonants and vowels differently, and so investigate multiplication (repeated addition) and division (the reverse of multiplication) by 10. Students in Years 3 and 4 should find they can tackle Level 3, but you will undoubtedly have students outside these ranges who could do these activities as an extension exercise.

Table 1.2: Australian Curriculum content descriptions for the *Numbers and words* activity

Activity level	Problem	Content descriptions
1	3- and 4-letter words	*Foundation* Connect number names, numerals and quantities, including zero, initially up to 10 and then beyond (ACMNA002) *Year 1* Represent and solve simple addition and subtraction problems using a range of strategies including counting on, partitioning and rearranging parts (ACMNA015) *Year 2* Solve simple addition and subtraction problems using a range of efficient mental and written strategies (ACMNA030)
2	Your name	*Year 1* ACMNA015 (see above) *Year 2* ACMNA030 (see above) *Year 3* Recall addition facts for single-digit numbers and related subtraction facts to develop increasingly efficient mental strategies for computation (ACMNA055)
3	Vowels are 5	*Year 3* Recall multiplication facts of two, three, five and ten and related division facts (ACMNA056) Represent and solve problems involving multiplication using efficient mental and written strategies and appropriate digital technologies (ACMNA057) Model and represent unit fractions including ½, ¼, ⅓, ⅕ and their multiples to a complete whole (ACMNA058) *Year 4* Investigate number sequences involving multiples of 3, 4, 6, 7, 8 and 9 (ACMNA074) Recall multiplication facts up to 10 × 10 and related division facts (ACMNA075)

The *Numbers and words* activity links literacy and numeracy, and provides a chance to practise addition and vocabulary in a novel situation. In the initial problem we look at changing words into numbers and finding the sum of those numbers, which we call the score of the word. We vary this idea in a number of ways. For example, we consider the scores of the numbers that are assigned in different ways to the letters of words; we look at biggest and smallest sums; we look at when two numbers have the same score; and we look at division by 10. You may want to develop the problem over a series of weeks to follow different ideas and ranges of numbers.

The basic idea of assigning letters to words has been used for a long time to encode messages. Apart from spy movies, codes are used on an everyday basis for sending material that needs to be kept secret. For example, codes are used between financial institutions, for mobile phones, government communications and so on. You might note that converting letters to numbers is easier today, as computers can handle the calculations involved. One such code, the RSA cryptosystem, is based on very large prime numbers.

Big ideas

» Use the four processes of addition, subtraction, multiplication and division with numbers less than 100 and less than 1000
» Largest and smallest scores (place value)

Problem aims

» Interpret and solve word problems
» Think logically
» Extend the problem-solving strategies
» Generalise the outcomes from the solution

Key concepts

» Understand the relation between various parts of a problem in order to solve it
» Record and justify the answer using appropriate notation
» Use the basic arithmetic operations in a 'real' setting/context

Possible heuristics/strategies

» Trial and error
» Exhaustive search (try all possible cases)

Level 1: 3- and 4-letter words

Problem

Ask the students to draw a picture of a cat and write the word CAT underneath.

Link the letters of the alphabet to the numbers 1 to 26 as shown below. Display this table on the whiteboard or put it on a handout.

A	B	C	D	E	F	G	H	I	J	K	L	M
1	2	3	4	5	6	7	8	9	10	11	12	13
N	O	P	Q	R	S	T	U	V	W	X	Y	Z
14	15	16	17	18	19	20	21	22	23	24	25	26

Ask them to put the right numbers under the letters of CAT. What do the numbers of the word CAT add up to?

You might want to use one of the other tables given on page 10 for some or all of your class, or to start with the smaller numbers as an example. In the text below we will use the table with the numbers 1 to 26, but corresponding questions can be asked for the other tables.

Problem steps

Step 1

Tell your class the problem. Then get students to draw a cat and put the letters C, A, T underneath. Under C they should then put 3, under A put 1 and under the T put 20. Then ask them to do the addition and let them put the total 24 under or beside the numbers.

Step 2

To help students see the idea of the correspondence between letters and numbers, let them make more pictures of objects with three letters like DOG, BAG, TEA and so on. Ask them to put the right numbers below each letter. They then add up the numbers for those words to give their *score*.

Step 3

Give the class some appropriate pictures with numbers but not letters. They should put the right letters in place and do the addition.

Step 4

Let selected students write down the numbers for some 3-letter words, in each word's order. The rest of the students have to find the letters that they have encoded and draw the picture of that object. What do the numbers add up to?

Step 5

Let the students choose four more 3-letter words, decode and draw them. Ask them what these numbers add up to. Can they tell you which word they have used so far has the biggest sum? Ask which word they have used so far has the smallest sum.

Step 6

Now repeat Steps 2 to 5, but using 4-letter words.

Step 7

Can they find 3- and 4-letter words with a score bigger (or smaller) than any of the ones they have found so far?

Where to from here?

- What do they think was the hardest thing that they did in this activity?
- Ask your students: Which sporting club's name has the biggest score and which has the smallest score? Repeat with pets' names or any other names they like.
- Is it true that some numbers will not be the scores of a word? Is it true that some numbers' scores represent at least three words? Can the score of some 3-letter words be the same as the score of some 4-letter words?
- Ask your students for ideas that they have for other problems that are like the *Numbers and words* activity.

Level 2: Your name

Problem

Link the letters of the alphabet to the numbers 1 to 26 as shown below. Display this table on the whiteboard or put it on a handout.

A	B	C	D	E	F	G	H	I	J	K	L	M
1	2	3	4	5	6	7	8	9	10	11	12	13
N	O	P	Q	R	S	T	U	V	W	X	Y	Z
14	15	16	17	18	19	20	21	22	23	24	25	26

Get the students to write out their names. Ask them to convert the letters of their names to numbers using the table above. What score do the numbers of their name add up to?

Note that, as in Level 1, you don't have to use the 1 to 26 table.

Problem steps

Step 1

Get your students to put the scores of their names on the board. Who has the largest score? Who has the smallest score? Do any of your students have the same score? Why? Is this because they have the same name or letters, or is it a coincidence?

Step 2

Let the students choose which letters in the table are given the numbers 1 to 26 (or they can choose numbers they feel comfortable using). Put this new task on the board for everyone to see. Let them find their name's new score. Who has the biggest score now? Who has the smallest score? Do any students have the same score? Why?

Step 3

Let the students play *the joker*. This means they can change the number of any letter in their name to 50. What is the score of the letters of their name now? Who has the largest and smallest scores? What was the best strategy for assigning the numbers?

Discuss this with the whole class to see the different strategies. The best thing to do will depend on the name. Sometimes changing the smallest number in their name to 50 will be best—but if a letter in their name is repeated, it may be best to apply the joker to those letters. Each name may have its own best strategy. Discuss the best approach with specific names.

Step 4

Using the same assignment of numbers to letters as in Step 2, each student chooses a particular letter in their name. Then they add up all of the other numbers and *subtract* the chosen letter/number. Which number do they think gives them their biggest score? Can they get the number that is closest to zero?

Discuss the answers with the whole class. Does anyone want to change their choice of letter? Is there a good strategy here? The best strategy will vary, but it may be that choosing the smallest letter will be best.

Step 5

Now let students assign the numbers to letters in any way that they want.

Repeat Step 4, so that each student chooses a letter in their name. Let them subtract the value of this letter from the sum of the other letters. Can they assign the numbers so that they get a sum of zero? Can anyone do this in more than one way?

Discuss who was able to do this and who wasn't. Is there a strategy? (General strategy: give numbers to all letters but one, then add these numbers and assign the total number to the final letter. Alternatively, choose the letter to subtract and give that letter any number you like. Then give the other letters numbers that add to the chosen letter's value.)

Where to from here?

- Ask them what was the hardest thing that they did for this task.
- What did they think was the best way to choose which letter was to be made the joker?
- What was the best strategy to use in the subtraction problems?
- What problems can they invent that are like the name problems here?

Level 3: Vowels are 5

Problem

Let consonants be worth 1 and vowels (A, E, I, O, U) be worth 5. Can your students find what the word TEA is worth if you have to *multiply* the value of each letter together?

Again, if you think that this is not suitable for all your students you could change 5 to 2 or even 10. More able students might be challenged by other numbers.

Problem steps

Step 1

TEA is worth $1 \times 5 \times 5 = 25$.

Ask your students: If vowels are worth 5 and consonants 1, what possible scores could they get for a 3-letter word?

- If a word has no vowels, they will get a score of 1.
- If there is only one vowel, they will get a score of 5.
- If there are two vowels they will get a score of 25.
- If they are all vowels they will get a score of 125.

Step 2

Ask the students to find four or more 3-letter words that each have a score of 1, 5, 25 and 125.

Here are some examples:

- Scores of 1: fly, sly, sty, try. But there aren't many words that have no vowels. (If we disallow the letter 'y', can this problem be solved?)
- Scores of 5: bat, cat, dot, paw. There are a lot of these words.
- Scores of 25: ape, bee, eat, pea. (If we disallow the letter 'e', can you solve this problem?)
- Scores of 125: We don't think that there are any English 3-letter words without a consonant. Can you find any in languages other than English? (In French, we can find 'eau' and 'oui'.)

Step 3

Ask the students: Why do these scores go up in multiples of 5? (If you replace a 1 by a 5 the score is multiplied by 5.)

Step 4

Now let each student look at their first name. If vowels are 5 and consonants 1, what is their name worth? Whose name has the largest score? Whose name has the smallest score?

Step 5

Now let consonants be 2 and vowels 5. Choose any word. Multiply the numbers of the letters together. Can students find three words whose score is 200?

(Words have to have three consonants and two vowels; such words are very common.)

Step 6

Following on from Step 5, ask students to think of a 4-letter word and find its score. Now *divide* this score by 10. What is the smallest answer they can get?

(Any word with three consonants and one vowel will give you an answer of 4. There is a small number of 4-letter words with no vowels, such as *hymn, myth* and *lynx*, which give an answer of 1.6. Most students at these year levels are not likely to know these words, but be ready in case they surprise you.)

Step 7

Choose any 4-letter word. What do students have to divide it by to get an *answer* of 10? Can you find a number to divide this word's score by to get answers of 50, 25, 8, 5, 4, 1, $\frac{1}{2}$, $\frac{1}{5}$, $\frac{1}{10}$, $\frac{1}{20}$, $\frac{1}{50}$? Get some students to fill in the tables below for their words on the board. Can you find a divisor for all words?

A word with three consonants and one vowel will give the results in Table 1.3.

Table 1.3: Conversion of letters to numbers

Answer	50	25	10	8	5	4	1	$\frac{1}{2}$	$\frac{1}{5}$	$\frac{1}{10}$	$\frac{1}{20}$	$\frac{1}{50}$
Divisor	no*	no*	4	5	8	10	40	80	200	400	800	2000

(*Although we say there is no divisor for 50 or 25, this is not technically true: $\frac{4}{5}$ is a divisor for 50 and $\frac{8}{5}$ a divisor for 25, but few students are likely to be able to see this.)

A word with two consonants and two vowels will give the results shown in Table 1.4.

Table 1.4: Conversion of letters to numbers

Answer	50	25	10	8	5	4	1	$\frac{1}{2}$	$\frac{1}{5}$	$\frac{1}{10}$	$\frac{1}{20}$	$\frac{1}{50}$
Divisor	2	4	10	no*	20	25	100	200	500	1000	2000	5000

(*Again, there is a divisor – $\frac{8}{100}$ or $\frac{2}{25}$ – but your students are unlikely to understand this.)

Discuss these results with the class, paying special attention to the answers that are not possible.

Where to from here?

- What were the easiest problems here?
- What possible fractions can they get for the division problem?
- What happens if you use different numbers for vowels and consonants?
- It might be interesting for your students to investigate coding methods and to practise them on other members of the class. Let them start with the Caesar cipher, which you can find via the series website and then make up their own codes.
- What problems can your students invent along the lines of the problems here?

CHAPTER 2: CATS AND DOGS

Initial problem

It is said that each year a camel lives is equivalent to two human years. Ask your students: what does this mean? How can they find the 'human age' of a camel?

Background information

This problem is based on a question that comes up from time to time in children's conversations: *how old is my cat or dog in people years?* This might arise when they want to know if their pet is equivalent to the age of their sibling, parent or grandparent.

The emphasis in this problem is on addition, multiplication and constructing tables and graphs, in relation to the relative ages of animals and people.

In Level 1 we use a 2-to-1 equivalence for camels and essentially construct a multiplication table using repeated addition. We repeat the camels' activity with cats, where there is roughly a 5-to-1 equivalence. Although cats are more familiar to your students, we use camels first as the number 2 may be easier for them to deal with than the 5 for cats.

These Level 1 activities are suitable for students at all levels. For Foundation and Year 1 students it might be better if they did the camels part first, because the numbers are smaller and they will find doubling easier, but there are not too many calculations involved. For older students the camels part could be done more independently, as an extension of what they have done with cats.

Level 2 does something similar to Level 1, but with dogs and lorikeets; here the respective numbers are 7 and 4. Level 2 is suitable for students from Year 2 upward, and also some Year 1 students; it is probably not suitable for most Foundation students. However, many steps are very close to the steps in Level 1, and these could be considered for most students at the younger year levels. Again, if you change the order so that lorikeets come first, the problem may be accessible to more students as 4 is easier for them to deal with than 7.

At Level 3 we look at the conversion of cat ages more carefully, and construct pictograms or graphs to make comparisons between the numbers at this level and with those in Level 1. Level 3 is more appropriate for Year 3 and 4 students; however, many Level 2 students should be able to cope with this.

This activity can be used to introduce and/or develop multiplication as well as to draw tables of values (Levels 1 and 2) and pictograms (all levels). Level 3 looks at what lies behind the numbers and the reason for thinking about this problem in the first place.

Students who have some computer knowledge and skills could potentially use the graph-making program on Microsoft Excel, especially at Level 3.

Table 1.5: Australian Curriculum content descriptions for the *Cats and dogs* activity

Activity level	Problem	Content descriptions
1	Camels and cats	*Foundation* Represent practical situations to model addition and sharing (ACMNA004) *Year 1* Represent and solve simple addition and subtraction problems using a range of strategies including counting on, partitioning and rearranging parts (ACMNA015) *Year 2* Solve simple addition and subtraction problems using a range of efficient mental and written strategies (ACMNA030)
2	Dogs and lorikeets	*Year 1* ACMNA015 (see above) *Year 2* ACMNA030 (see above) Recognise and represent multiplication as repeated addition, groups and arrays (ACMNA031) *Year 3* Recall addition facts for single-digit numbers and related subtraction facts to develop increasingly efficient mental strategies for computation (ACMNA055) Represent and solve problems involving multiplication using efficient mental and written strategies and appropriate digital technologies (ACMNA057)

Table 1.5: Australian Curriculum content descriptions for the *Cats and dogs* activity (continued)

Activity level	Problem	Content descriptions
3	Age graphs	*Year 1* Represent data with objects and drawings where one object or drawing represents one data value. Describe the displays (ACMSP263) Create displays of data using lists, table and picture graphs and interpret them (ACMSP050) Collect data, organise into categories and create displays using lists, tables, picture graphs and simple column graphs, with and without the use of digital technologies (ACMSP069) Interpret and compare data displays (ACMSP070) *Year 2* ACMNA030 (see above) ACMNA031 (see above) ACMSP050 (see above) *Year 3* ACMNA055 (see above) ACMNA057 (see above) ACMSP069 (see above) ACMSP070 (see above) *Year 4* Investigate number sequences involving multiples of 3, 4, 6, 7, 8 and 9 (ACMNA074) Develop efficient mental and written strategies and use appropriate digital technologies for multiplication and for division where there is no remainder (ACMNA076) Construct suitable data displays, with and without the use of digital technologies, from given or collected data. Include tables, column graphs and picture graphs where one picture can represent many data values (ACMSP096)

Note: All the questions asked in the text are directed to the students.

Big ideas

» Use addition and multiplication
» Use tables and graphs to represent data and show that data can be represented in more than one way
» Understand what graphs can tell us

Suggested resources
- Butcher's paper
- Blank tables
- Graph paper

Problem aims
- Develop mathematical language involving comparisons
- Discover why multiplication is useful in everyday life
- Understand the process of multiplication
- Interpret and solve problems in a real-world situation
- Construct tables and pictograms/graphs
- Interpret data from a table or pictogram/graph

Key concepts
- Basic multiplication
- Translating data into tables and graphs
- Reading tables and graphs with understanding

Possible heuristics/strategies
- Trial and error
- Trial and improve
- Make a table
- Make a pictogram
- Draw a graph (by hand or by using Microsoft Excel)

Level 1: Camels and cats

Problem

It is said that each year a camel lives is equivalent to two human years. Ask your students: What does this mean? How can they find the 'human age' of a camel?

Prior to the lesson: Ask your students to bring a photo or picture of a camel to school, whose age they will have to guess. It would also be good for them to bring a photo of two members of their family and give their ages. Family members' ages should be given in multiples of 2 or 5—so if they bring a photo of their 67-year-old grandfather, they should say his age is 66 or 65.

Problem steps

Step 1

What does the question mean? How can a camel year be equivalent to two human years?

In this activity we mean that a camel that is one year old has developed to approximately the stage of a two-year-old human. A camel that is two years old has developed to approximately the stage of a four-year-old human, and so on. For ease of calculation, we assume that camels live for 50 years and humans live for 100 years.

(There are some big assumptions here; it would be useful to discuss them with your students.)

Ask your students how many 'actual' years old they think the camel in their picture is. (You may want to ask this the day you start the activity, so that they will have the answer ready when you need it.) So how old does that make their camel in people years?

Step 2

Ask your students how they can make the calculations for the comparisons between camels and people easier. What possibilities can they think of? Discuss using a calculator, making a table, drawing a graph and so on.

Step 3

Make a class table like the one in Table 1.6. Use butcher's paper or a whiteboard so that the table is large enough for everyone to see. Draw up the rows and columns before the class starts and add the first two column headings; add the other headings as the lesson progresses.

First fill in Column 1 and one or two entries from Column 2. Talk to the students about how you go from one year to the next in each column.

Table 1.6: Conversion of camel ages to people ages (entries in columns 3 and 4 are examples)

Camel age	People age	Camel pictures	Some people we know
1	2		Cheng's younger brother
2	4	Tom's camel	
3	6		
4	8		
5	10		
6	12		Michael's brother
7	14		
8	16		
9	18	Eli's camel	
10	20		
11	22		Sue's cousin
12	24		
13	26		
14	28		Kwai's aunt
15	30		
16	32		Australian cricket captain
17	34		
18	36		
19	38		Ding Jung's uncle
20	40		

Then get the students to fill in Column 3 with their camel pictures. Talk about how old the camels in the pictures are. Can they tell if the camel is young, middle-aged, old or very old? Whose camel is the oldest? Whose camel is the youngest? Is Mary's camel older than Eli's?

Finally, use the ages of people that they know to get a feel for what the numbers in Column 1 mean. It might be useful to repeat the questions from the last paragraph again here.

Step 4

How else could your students represent this data? If possible, have them construct a pictogram like the one in Figure 1.1 over the page. Note that the camels are twice as tall as the humans in the pictogram because one year for camels is equivalent to two years of a human life.

Figure 1.1: A pictogram for camel and people ages

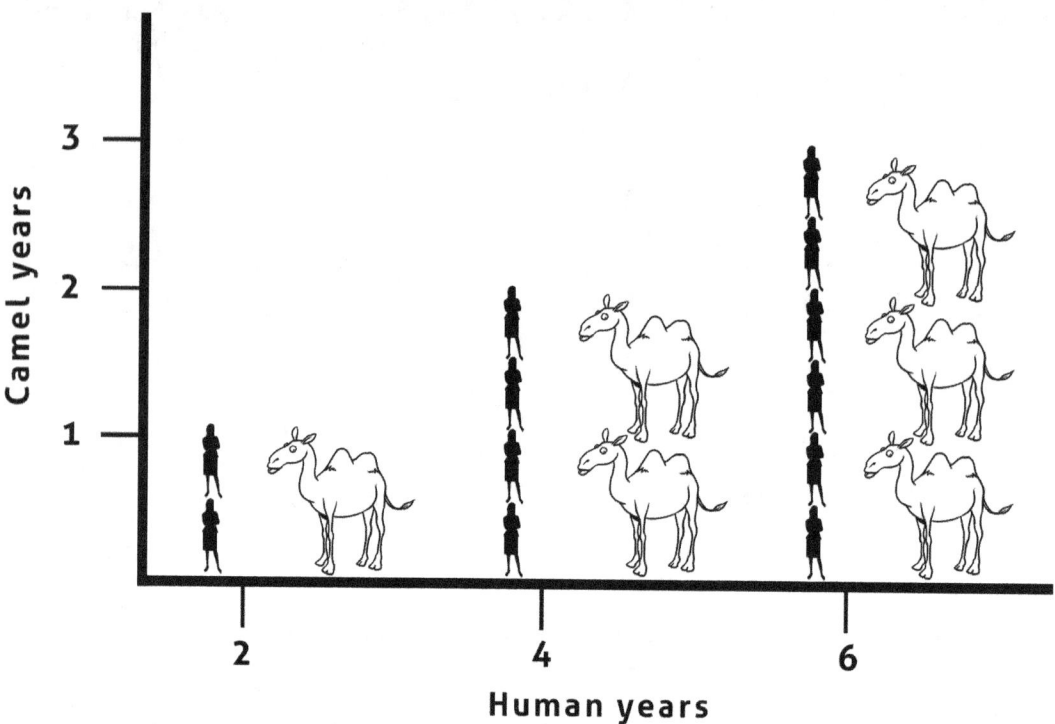

Step 5

Now let's look at cat ages. Suppose that cats live to be 20. (We have chosen the number 20 so that students can be led to a possible conversion factor of 5.) See, for example, Table 1.7.

Ask students how they can change cat years into people years. Can they make up an age conversion table for cats?

Table 1.7: Cat ages to people ages

Cat age	People age	Cat age	People age
1	5	11	55
2	10	12	60
3	15	13	65
4	20	14	70
5	25	15	75
6	30	16	80
7	35	17	85
8	40	18	90
9	45	19	95
10	50	20	100

From this point, the activity repeats what has already been done with camels' ages, so you may want the students to do the 'cats' work more independently than they did the 'camels' work.

Get your students to draw and/or complete a table like the camel conversion one. The cats' ages should vary from 1 to 20. (There is a pre-made table in the student handouts available on the series website.) For visual engagement, give the students some cat pictures (you can find plenty online) or let them draw a few cats of all ages. They could also make a pictogram similar to the one in Figure 1.2 or even add cats to the pictogram in Figure 1.1.

Figure 1.2: A pictogram for cat and people ages

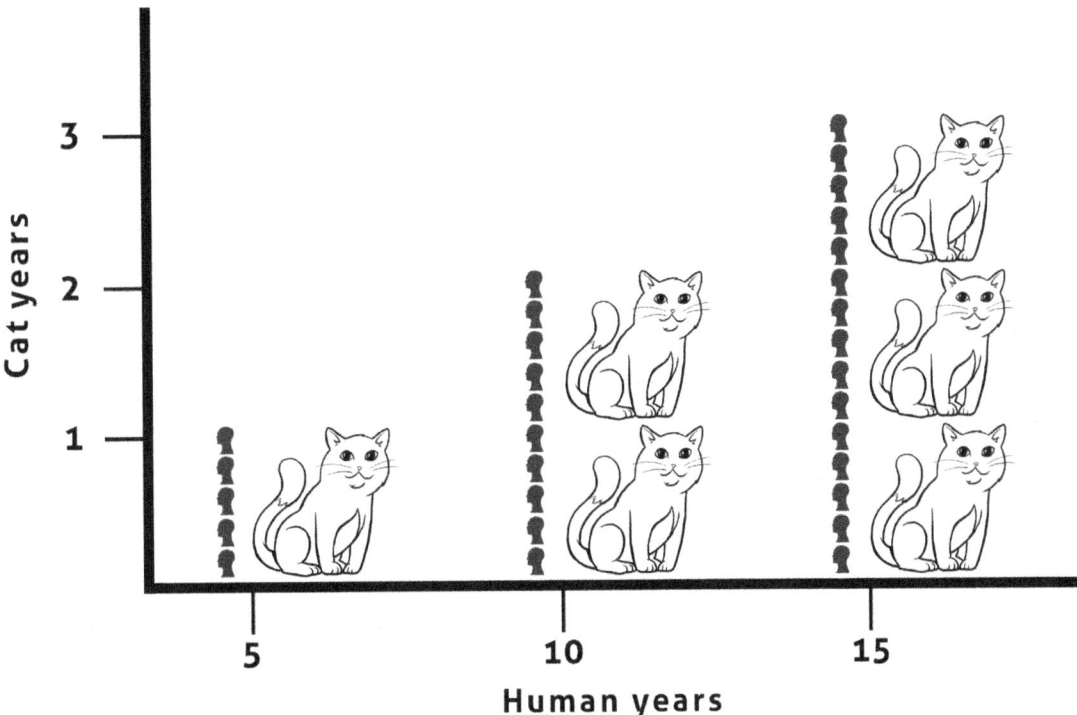

Where to from here?

- What do they think was the hardest sum they had to do here? Did the table help them to do the calculations?
- Would you rather be a cat or a camel? Why?
- Can you make up a problem like the ones above?

Level 2: Dogs and lorikeets

Problem

It is sometimes said that each year a dog lives is equivalent to seven human years. So how old are your students' (or their friends') dogs in people years? What do they think that this means?

Note: Lorikeets have a simpler conversion number, so you may like to do those steps before you look at dogs.

Prior to the lesson: Ask those students with dogs to bring a photo of them to school, and for all students to bring back the family member photos from the Level 1 activity.

Problem outline

Step 1

Discuss the problem so that they understand what they are about to do. Remind them of the Level 1 problems and explain that this is a similar activity.

Step 2

Ask students how they can find the approximate human age of their dog.

Multiplication or repeated addition is one possible option. What strategy do your students use to find the age of their dog (or a friend's dog)? Think of how you might support students who struggle with multiplication.

Step 3

Discuss the results of the previous step. This is easiest done by drawing up a table like Table 1.6 from Level 1. (It's also easier to do the calculations this way.)

Table 1.8: Dog/human years conversion table (entries in columns 3 and 4 are examples)

Dog age	People age	Dog pictures	Some people we know
1	7		Cheng's brother
2	14	Tom's dog	
3	21		
4	28		
5	35		Australian rugby league captain
6	42		Matilda's father
7	49		
8	56		Prime minister
9	63	Eli's dog	
10	70		
11	77		Gerhan's grandfather
12	84		
13	91		
14	98		Pierre's great grandmother
15	105		
16	112		
17	119		

Have your students add their dog and people photos to fill out the table. How old are their dogs in people years? Are the dogs in the photos young, middle-aged, old or very old? Who has the youngest puppy and the oldest dog in your class?

Can they use the ages of people that they know to get a feel for what the numbers mean? (Many children think that 30 is 'very old'.)

Step 4

Can the students tell you how old the people in Table 1.8 are in dog maturity years?

For example, a person is as mature after 14 calendar years as a dog is after two years. So what would a one-year-old baby be equivalent to in dog maturity?

Now get them to make a table of the age of people in dog years. This requires them to be able to use Table 1.8 'backwards' with some idea of division.

Step 5

Some lorikeets live to be 25 years old. Get your students to draw up an age comparison table for lorikeets. (This could resemble Table 1.8, but without columns 3 and 4.)

What conversion factor will they use? To keep things in round numbers, it's simplest to say that 25 lorikeet years is equivalent to 100 human years.

Step 6

Ask your students to work out from the table how old the members of their family would be if they were lorikeets.

Step 7

The Life Span of Animals website gives the life expectancy for a range of animals. What conversion factors can the students find for different animals? (These may be rounded to whole numbers, though if some children can handle fractions or decimals they can give a more exact equivalent.)

To connect with other class work, you might have the students produce tables where the factor is a number you've been working with recently. To do this, ask them to find animals whose conversion factor is that specific number.

Where to from here?

- Ask your students about the disadvantages of being a dog compared to a lorikeet.
- Would your students prefer to be cats, dogs, camels or lorikeets? Why?
- What are the advantages of using a table?
- Ask your class how accurate their animal conversion charts are. Let them explore online to find out.
- Can your students think of any questions based on the ideas of this level?

Level 3: Age graphs

Problem

One cat-to-human age chart found online, shown here in Table 1.9, gives a more realistic conversion of cat ages to people ages.

Table 1.9: More realistic cat ages to people ages

Cat age	People age
1	15
2	24
3	28
4	32
5	36
6	40
7	44
8	48
9	52
10	56
11	60
12	64
13	68
14	72
15	76
16	80
17	84
18	88
19	92
20	96

What patterns can your students see in the numbers here? Provide help where necessary.

Problem steps

Step 1

How can your class make a picture of this? There are at least two ways to do this. One is using a pictogram and another by using a graph.

(We show the graph method here, but the pictogram may be more accessible to your Year 3 and 4 students. The questions that we use for graphs have parallels in pictograms. If they use a pictogram you may want to reduce the cats' data so that it fits the paper available more exactly.)

One way to draw a graph of the data in Table 1.9 above is to use Microsoft Excel or some other spreadsheet program. We show the graph from one of these programs in Figure 1.3. However, it can also be done by plotting points. For instance, a cat aged 10 years is equivalent in age to a person aged 56 years. So go along 10 units on the horizontal axis and go up 56 units on the vertical axis and put a dot or some other small mark. Do this for all of the data in Table 1.9.

Figure 1.3: A graph of cat ages drawn from web data

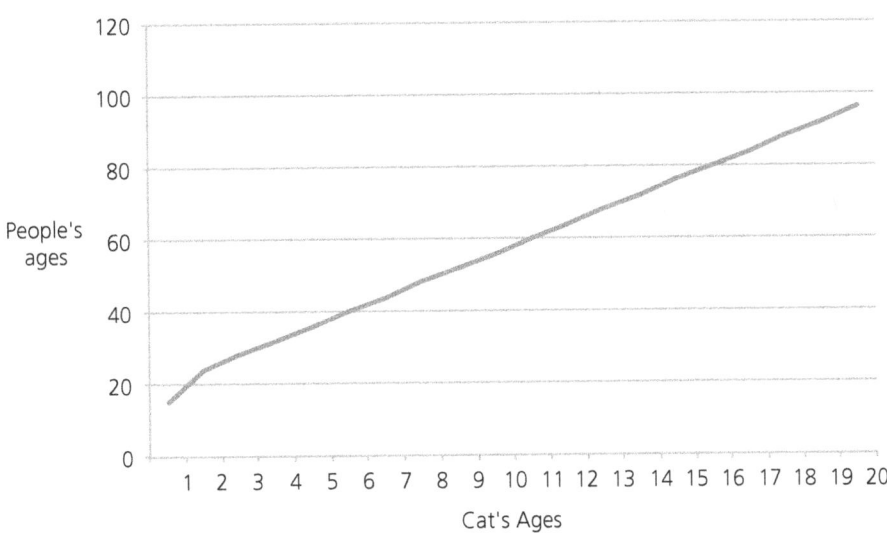

Step 2

Check that the students understand how to use their graph by asking questions such as: What is the human age of a cat that is 9? What is the cat age of a human who is 70? What is the human age of a cat that has lived as long as they have?

Get the students to ask similar questions of each other.

Step 3

Ask students to make a pictogram or graph of the cat data from Table 1.7 in Level 1 (see Figure 1.4).

Figure 1.4: A graph of cat ages drawn from Table 1.7

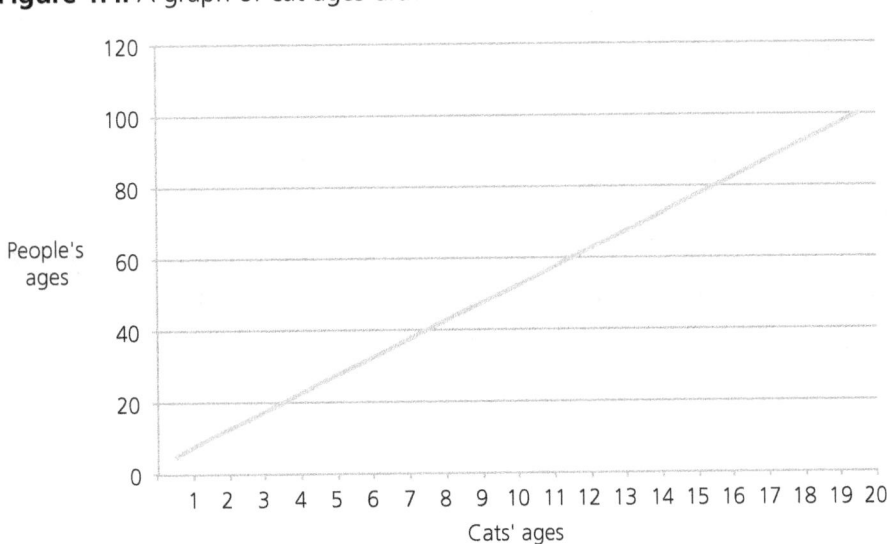

Step 4

Check that the students understand how to use their pictograms or the graph of Figure 1.4 by asking questions such as: What is the human age of a cat that is 10? What is the cat age of a human who is 70?

Get students to ask similar questions of each other.

Do the students get the same answers from Figure 1.4 as they do from Figure 1.3?

Step 5

Ask the class to put their graphs of figures 1.3 and 1.4 together (Figure 1.5) to make it easier to compare the different information from the tables that were used to make them.

Figure 1.5: A graph of cat ages comparing two types of data

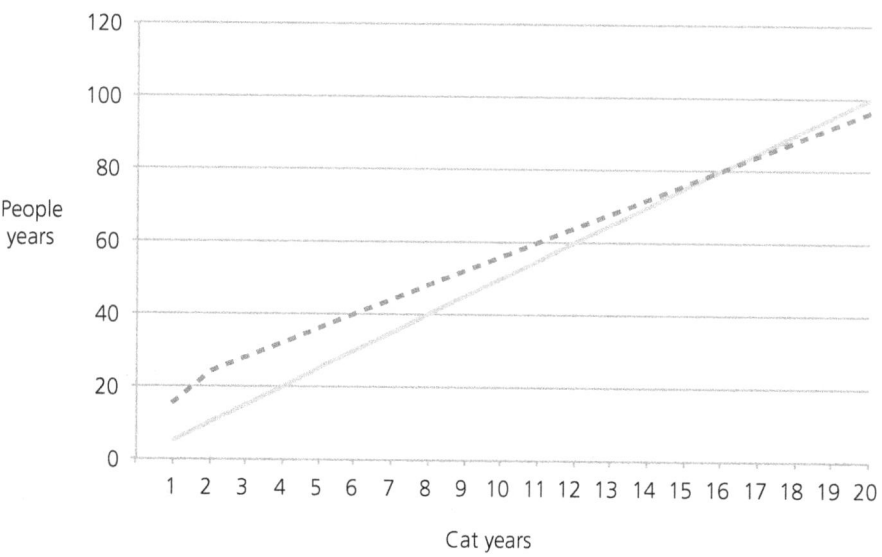

Are the graphs very different? Discuss whether it is easier to see any difference using the tables or the graphs.

The two graphs are only the same when the cat's age is about 16 years. In the region between ages 11 and 20, though, there is not very much difference. The solid line is straight; the dashed line has two straight pieces. For most of its length, the solid line is steeper than the dashed line. Does this mean that cats age more quickly in human terms using the graph with the solid line or the dashed line?

Step 6

Ask the class: Why would anyone want to know a cat's age in people years?

We suspect that this comparison of ages is just a rule of thumb that people use to get a feel for how their own ages and their cats' ages compare. It might not be of any great practical use. Do veterinarians or anyone else use it for any practical purpose?

Step 7

Look up more realistic data for dogs and cats online. Ask your students to draw graphs for this data.

Get the students to compare their camel and dog graphs. What do the graphs tell them? (They should tell you that a dog ages faster than a camel etc.)

Where to from here?

- Ask your class if they prefer to work with multiplication (using the conversion factor), tables or graphs. What are the advantages and disadvantages of each of these?
- Can your students find other uses for tables and graphs?
- Invent some problems using graphs.

What was the cat data of Figure 1.3 based on? The Calculator Cat website has another set of data. For comparison and reference, the *Cat Bible*, a book by Tracie Hotchner, provides the following list:

- 1-month-old kitten = 6-month-old human baby
- 3-month-old kitten = 4-year-old child
- 6-month-old kitten = 10 human years old
- 8-month-old kitten = 15-year-old human
- A 1-year-old cat has reached adulthood, the equivalent of 18 human years
- 2 human years = 24 cat years
- 4 human years = 35 cat years
- 6 human years = 42 cat years
- 8 human years = 50 cat years
- 10 human years = 60 cat years
- 12 human years = 70 cat years
- 14 human years = 80 cat years
- 16 human years = 84 cat years

So this gives some idea of how the data is found. But the data here differs from the data in Table 1.9. Ask students where the numbers differ. How accurately do we need to know a cat's age in human terms?

- The Life Span of Animals website gives the life expectancy for a range of animals. What conversion factors can you find for different animals?
- You might want your students to produce tables where the factor is a number that you have been working with in class. In that case, you could ask them to find animals whose conversion factor is that specific number.
- How accurate are the animal conversion charts? Let them explore online to find out.

CHAPTER 3: DIFFERENCE AND SUM PROBLEMS

Initial problem

Hamish and Claire were playing around with 2-digit numbers.

Hamish did this calculation: 64 − 46 = 18. He then added 18 to its reverse, 81, and got 99.

Claire thought that was strange because she had taken 93, reversed it to get 39 and then took 39 from 93. This gave her 54. Then she added 54 to its reverse, 45. She got 99 too.

Claire wondered (*conjectured*) if you always get 99 this way. What do your students think?

Background information

This activity involves differences and sums (Table 1.10). It's important to note that, in its present form, this problem is probably too difficult for most Foundation, Year 1 and Year 2 students, though most Year 3 and Year 4 students should be successful.

Let's pause a minute. First of all, a pattern is a pattern. Things like 2, 4, 6, 8, 10 …; 3, 6, 12, 24, 48 …; and 5, 25, 125, 15 625, 244 140 625 … are all patterns. But mathematicians think that some patterns are nicer than others, so they call them 'nice' or 'pretty' or even 'beautiful'.

They stand out as different from other patterns either because they take old patterns that are familiar and add a little twist, or because they are totally new or unexpected. For example, in Level 1, Step 3, there is a pattern in the *vers* – they are always multiples of 9. You can think of this as 'nice' if you like (the property of 'niceness' is in the eye of the beholder), but the fact that a *rev* can only be 0 or 99 gives a pattern that you might think is nicer than the *vers'* pattern. There are only two numbers in the pattern and it is likely to be unexpected because it is not at all obvious that anything so neat would occur. We encourage your students to think about niceness in maths, both for patterns and for other things such as ideas. Someone may produce an idea that is totally unexpected and useful in a given situation.

However, there are two reasons why this problem is in this book. First, because there are some 'nice' patterns here, and second, it gives a chance to practise addition and subtraction of 2-, 3- and 4-digit numbers in a situation where the point is to discover a pattern.

Because of the value in searching for and finding patterns, as well as preparing students to investigate addition and subtraction of numbers, we recommend that students are allowed to use calculators here. This will allow more Year 1 and all Year 2 students to participate in this activity.

The overall activity is based on the following process:

» take a number
» reverse its digits
» subtract the smaller of the two from the other to make a *ver* (or whatever name your students would like to give it)
» reverse the digits
» add the result and its reverse to give a *rev*.

Level 1 concentrates on 2-digit numbers with a search for patterns in the *vers* that leads to the pattern in the *revs*. We look at the different results that can be created from the original problem, and along the way look at the first part of the calculation to see what patterns the subtraction gives. Level 1 is mainly for Year 2 students and above. Some students who find the calculations hard to do might be able to contribute to the conjecture discussion.

Level 2 does the same thing with 3-digit numbers. At this level we also investigate proofs of conjectures from Level 1. Level 2 is available for many Year 2 and Year 3 students and above.

Level 3 concentrates on the *vers* and *revs* of 4-digit numbers, and looks at proofs for open questions from Level 2. We also experiment with 4-digits and see that things are not as they would seem from Levels 1 and 2. Level 3 is probably best left for Years 3 and 4 students, though parts of it could be considered by Year 2 students if they used calculators.

Table 1.10: Australian Curriculum content descriptions for the *Difference and sum* activity

Activity level	Problem	Content descriptions
1	2-digit numbers	*Year 1* Count collections to 100 by partitioning numbers using place value (ACMNA014) *Year 2* Investigate number sequences, initially those increasing and decreasing by twos, threes, fives and ten from any starting point, then moving to other sequences (ACMNA026) Solve simple addition and subtraction problems using a range of efficient mental and written strategies (ACMNA030)
2	3-digit numbers	*Year 1* ACMNA014 (see above) *Year 2* ACMNA026 (see above) ACMNA030 (see above) *Year 3* Apply place value to partition, rearrange and regroup numbers to at least 10 000 to assist calculations and solve problems (ACMNA053)
3	4-digit numbers	*Year 2* ACMNA030 (see above) *Year 3* ACMNA053 (see above) *Year 4* Apply place value to partition, rearrange and regroup numbers to at least tens of thousands to assist calculations and solve problems (ACMNA073) Investigate number sequences involving multiples of 3, 4, 6, 7, 8 and 9 (ACMNA074)

There are some unexpected things going on here. What happens and why is not clear-cut and sometimes all we can do is to end with a conjecture—a guess as to why the numbers are behaving the way they do. But we're doing this for two reasons. First, it shows that patterns don't go on and on as we might sometimes think. Second, it shows that there are some problems that we have to give up on because they just become too hard at some point. This mirrors what happens when mathematicians are working; they aren't able to solve all of the problems they try to solve either.

But it's also almost certain that proving any of the conjectures will be too much for most students. Some students may be able to get on top of the proof by exhaustion (looking at all possibilities) for the 2-digit case, and a few will be able to understand the other two approaches in Levels 2 and 3. But very few students will be able to get to the more sophisticated proof of the 3-digit case in Level 3.

This activity is based around a classic 'old chestnut', the 1089 Problem, which is presented as the Level 2 problem here. This problem can be turned into a magical trick; you can find it online via the series website.

One thing this activity shows is that you can take a well-known problem and play with it a bit to produce at least two other related problems; thus demonstrating that mathematics develops by extending known results.

Big ideas
» Conjecture patterns
» Prove/justify
» Use place value especially to add and subtract 2-, 3- and 4-digit numbers

Suggested resources
» Calculator

Problem aims
» To give practice in adding and subtracting numbers with 2- and 3-digit numbers in a problem situation
» To justify a result

Key concepts
» Understand a problem in order to solve it
» Discover that factors of 9 arise from the subtraction of certain 2- and 3-digit numbers

Possible heuristics/strategies
» Trial and error
» Exhaustive search (try all possible cases)
» Special cases

Special notes

Palindromes: A palindrome is a number that is the same whether it is written forwards or backwards. For instance, 171, 123 321 and 983 050 389 are all palindromes. On the other hand, 567, 398 931 and 999 099 are not palindromes.

Proof by exhaustion: This is a proof where you break the problem up into cases and show that each case works. The 'exhaustion' here refers to exhausting all the cases (not the person doing the proof). This is a simple way of proving problems and it can be used in a number of places.

Division by 9: There is a simple test for division by 9; when the sum of the digits of a number is divisible by 9, then so is the number. For example, for 18, 1 + 8 = 9 and 9 is a factor of 18; for 2448, 2 + 4 + 4 + 8 = 18, which is divisible by 9 and we can check that 2448 is divisible by 9. It works the other way too. If a number is divisible by 9 then so is the sum of its digits. Now 36 is divisible by 9 and 3 + 6 = 9; and 13 437 is divisible by 9 and 1 + 3 + 4 + 3 + 7 = 18 which is divisible by 9.

Chapter 3: Difference and sum problems

Level 1: 2-digit numbers

Problem

Hamish and Claire were playing around with 2-digit numbers.

Hamish did this calculation: 64 − 46 = 18. He then added 18 to its reverse, 81, and got 99.

Claire thought that that was strange because she had taken 93, reversed it to get 39 and then took 39 from 93. This gave her 54. Then she added 54 to its reverse, 45. She got 99 too.

Claire wondered (*conjectured*) if you always get 99 this way. What do your students think?

Problem steps

Step 1

Start the problem by getting one or two students to the board to be Claire or Hamish. Get these students to do a few examples to see if they end up with 99. For students using calculators, one student should do the calculation and the other write the answer on the board. Then ask the class what they could conjecture from these calculations. Are their conjectures the same as Claire's?

> **Possible class conjecture 1:** As a result of Claire and Hamish's process we end up with 0 or 99.

From here this problem needs some experimentation. Get your students to choose any 2-digit number they want, reverse it and subtract the smallest from the largest. Whatever the result, add that number to its reverse. Do they always get 99? (Nearly always.)

There is a problem with numbers like 65; 65 − 56 = 9. The number 9 isn't a 2-digit number so how can it be reversed? It can, if 9 is treated as 09. Reverse that to get 90. Then 90 + 09 = 99. In any situation when you get a single-digit answer and need to reverse its digits, treat it as a 2-digit number that starts with 0; for example, 6 becomes 06.

Step 2

It would be good for students to justify their conjectures, but that means that they should find some other patterns first. So let them concentrate on the subtraction part of Claire and Hamish's problem. It might help if they gave a name to the answer to this reversing and subtracting the smaller from the larger process. We'll call it a *ver*.

What patterns can they find in the *vers*?

Step 3

There are two patterns we'd like them to find, though they may find others.
1. All of the differences are multiples of 9.
2. Numbers where the two digits of the original number have the same difference give the same multiple of 9.

The first pattern says that the only answers you get for *vers* are 0, 9, 18, 27, 36, 45, 54, 63, 72, 81 and 90. The second says that 76 gives the same multiple of 9 as 10, 21, 32, 43, 54, 65, 87 and 98 (the difference between the digits is 1); so do 31, 42, 53, 64 (difference of 2) and so on.

Let the students work in small groups or alone to do a whole series of *ver* calculations. As they find answers, let them write them on the board. Stop when they can find no more. Then ask what all of these numbers have in common (they are multiples of 9).

Ask them next how they know when they have got all *vers*. For example, why don't they get 90 and 99? This might lead on to you asking what numbers, when *ver*-ed, give you 0 (palindromes, zero digit differences); 9 (numbers whose digit differences are 1); or 18 (digit differences of 2).

Can they make a conjecture? (If two numbers have the same digit difference, their *vers* are the same.)

Step 4

There is another pattern here that they may have met before. When you add the digits of a *ver* you either get 0 or 9. This is also true for 90 that isn't a *ver*. On the other hand, the digits of 99 add to 18 which is also a multiple of 9. It turns out that for every number that is divisible by 9, the sum of its digits is divisible by 9.

Lead the students, if necessary, to see this result for the *vers*.

Step 5

Now call the process of taking a number, reversing its digits and adding the two numbers together a *rev* (or whatever name your class would prefer). We now want to find the *rev* of a *ver*. If the *ver* is zero then so is its *rev*. So let's forget palindromes for a moment.

Take a *ver* that is not zero. You can move on from here in two ways. First you might get the class to find the *revs* of each *ver*. There are not too many and so they should be able to see that they only get 0 or 99, and the *possible class conjecture 1* is true. This method is called *proof by exhaustion*, because we have considered every case and exhausted all the possibilities.

On the other hand we know that its tens digit plus its units digit is 9. For example one possible *ver* is 81 and 8 + 1 = 9. Now try to find its *rev*.

$$\begin{array}{r} 81 \\ +18 \\ \hline 99 \end{array}$$

The *rev* is certainly 99. But there is something strange here. The *ver*, 81 this time, is 81 whether you read it across the sum above or down. The same thing is true for its reverse 18. So when you add the digits of 81 down you get the same as you do when you add them across. The same is true for 18. So the two numbers of the *rev* have to be 9 and 9. This shows that the *rev* is always 99. And we have proved the *possible class conjecture 1*.

Where to from here?

- Did your students appreciate the nice patterns here?
- What did your students think they needed to know to be able to solve this problem?
- What problems can they make up that are similar to this one?

Level 2: 3-digit numbers

Problem

Hamish and Claire were playing around with 3-digit numbers.

Hamish did this calculation: 614 – 416 = 198. He then added 198 to its reverse 891 and got 1089.

Claire thought that that was strange because she had taken 938 reversed it to get 839 and then subtracted 839 from 938. This gave her 99. Then she added 99 (= 099) to its reverse 990 and she got 1089 too.

Claire wondered (conjectured) if you always got 1089 this way. What do your students think?

Problem steps

Step 1

Let some students come out the front of the class to play the role of Claire and Hamish. Does the class have a better conjecture than Claire? Let the class experiment. On the basis of previous experience they should be wary of palindromes. Is your class happy with the conjecture below?

> **Possible class conjecture 2:** As a result of Claire and Hamish's process, they always end up with 0 or 1089.

Step 2

Give your students some time to work on the conjecture to make sure that they agree with it or are unable to find any *revs* other than 0 and 1089.

Step 3

As in Level 1, now investigate the non-zero *vers* and their properties. There are a large number of these properties that your students might find. We list some below.

- There are 9 *vers*.
- The *vers* are 99, 198, 297, 396, 495, 594, 693, 792, 891, 990.
- The tens digit is always 9.
- The unit and 100s digits add to 9.
- All the digits add up to 18.
- The sum of all the digits is divisible by 9 (see Special notes on p.37).
- The *vers* are the same as in Level 1 but with a 9 in the middle.
- All *vers* are divisible by 99.
- If two numbers have the same difference between their 100s and units digits they give the same *ver*.

Ask your students to justify as many of these properties as they can.

Step 4

In their groups, let students try to justify *possible class conjecture 2* by the two methods used in Level 1. We omit zero here as we know we can get that using palindromes. Different groups can be encouraged to explain their methods on the board.

First method: Find the *revs* of all of the *vers*. These will give them the answers they were expecting.

This is again a *proof by exhaustion*.

Second method: We give one example but all the *vers* obey this method because both the middle (tens digit) of a *ver* and the sum of the first and last numbers is 9.

$$\begin{array}{r} 891 \\ + 198 \\ \hline 1089 \end{array}$$

Because the tens addition gives you 18, the 1 of the 18 is a hundred and carries over to the hundreds column to give a *rev* of 1089.

Where to from here?

- Can your students explain all of the steps of the 3-digit problem? Which parts did they find most difficult?
- Ask your students to rate the niceness or otherwise of the patterns here.
- Ask your students where they used place value in this problem.
- Can they think up any other problems that look like the ones they have worked on here?

Level 3: 4-digit numbers

Problem

Hamish and Claire were playing around with 4-digit numbers.

Hamish did this calculation: 6124 − 4216 = 1908. He then added 1908 to its reverse 8091 and got 9999.

Claire thought that that was strange because she had taken 9358 reversed it to get 8539 and then subtracted 8539 from 9358. This gave her 0819. Then she added 0819 to its reverse 9180 and she got 9999 too.

Claire wondered (conjectured) if you always got 9999 this way. Ask your students what they think of this conjecture.

Problem steps

Step 1

This problem is quite difficult. It is not too hard to come up with a conjecture that is correct, but it is extremely difficult to prove the conjecture. It might be a good idea to let this one simmer in the background and come back to it from time to time.

If your class has already worked on the 2- and 3-digit cases, you might like to ask them to make a class conjecture straight away. It is likely that they will suggest that you can only get 0 and 9999.

Have two students come to the board and play the roles of Claire and Hamish. Does the work on the board justify the class conjecture or does it need to be altered? If a number other than 0 or 9999 comes up it will need to be checked. We have put the name for a conjecture below, but we don't know what your class might find.

> **Possible class conjecture 3:** Whatever your class decides.

Step 2

Get a few more volunteers to check what Claire and Hamish have done, but with different starting numbers. Is the conjecture looking good or does it need to be changed?

From there, get all the students to experiment to see what they can find out. As they find new ones, they should go to the board and add them to the 0 and 9999 that have been found so far. After a while you should see that a better conjecture is that the only possible answers are 0, 990, 9999, 10890, 10989. However, you should keep this information tight to your chest and only reveal it later.

Step 3

We proceed as we did in the previous levels by looking at *vers* before *revs*. After some experimenting, what can your class say about *revs*? What can they add to the list below?

- They are all a multiple of 9.
- Some are a multiple of 99, some a multiple of 999.
- The units digits are 0, 8 or 9.

- The tens digits are 0 or 9.
- Some of the *vers* are like those of Level 1 with 99 in the middle.

Step 4

Finding a good conjecture is hard enough but finding a proof is even harder. So we suggest that your students look at the *revs* of some particular numbers. These are 0000, 0001, 0010, 0011, 0100, 0101, 0110, 0111, 1000, 1001, 1010, 1100, 1101, 1110 and 1111.

Share these numbers among your class. Do these fill any gaps in the class conjecture? They should provide the class with all the possible *revs* listed in Step 2. So now revise the class conjecture.

> **Possible class conjecture 3 plus:** As a result of Claire and Hamish's process they will always end up with 0, 990, 9999, 10 890 and 10 989.

Step 5

The only thing that is worth doing now is to see if your students agree with the following conjecture.

- **Conjecture:** You get the same *rev* from two numbers if the difference between their 1000s and units digits are the same and if the difference between their 100s and tens digits are also the same.

  How does that help?

Where to from here?

- What were the key steps in the 2-digit and 3-digit justifications?
- Where did your students have to use place value in all of the levels of this activity?
- What do your students think was the 'nicest' thing they did in this level? What was the nicest argument they saw?
- Can they make up some problems that are similar to the 4-digit problem they have been working on in this level?
- Can your students guess what happens with 5-digit numbers?

CHAPTER 4: ARKS AND TARKS

Initial problem

Tarks are creatures from another solar system. Six Tarks are on the planet Eos. They want to get to the planet Helios by a space shuttle. Here are some things to bear in mind.

» Every space trip has to have one pilot.
» Every Tark is a qualified pilot.
» The shuttle can carry no more than four Tarks.

Make costumes for the Tarks and a model for the shuttle.

Is it possible to get everyone safely across to Helios?

Background information

Arks and Tarks involves problems about moving some aliens from one planet to another by shuttles. This is a logic problem, and students have to reason their way through it. It will be valuable for them to invent a notation to help them to keep track of where the aliens are after each shuttle flight.

Table 1.11 shows how the levels develop. At Level 1 there are six Tarks and the students have to get them from one planet to another in the least possible number of flights. The problem is gradually made more difficult by reducing the number of aliens who can travel on a shuttle at one time. The students are asked to find the least number of flights and, where possible, to justify this number. The problem develops by introducing another group of aliens, Arks, who may eat the Tarks in certain circumstances.

Level 1 considers moving six creatures from one planet to the next with a restriction on the shuttle size. At Level 2 we have to transfer three Arks and three Tarks and find the smallest number of flights needed, given an extra condition about the number of Arks and Tarks that can be together at any one time. Levels 1 and 2 are accessible to all students in the early years, as they involve little more than basic logic with some counting and looking for simple patterns. They can model the situation themselves or use plastic objects to model the Arks and Tarks.

At Level 3, we introduce a moon that can be used to help move larger numbers of aliens. Level 3 is not too much harder and should be appropriate for most students in the early years.

At both of the last two levels we investigate what happens with progressively smaller shuttles.

Some of the proofs in this activity may be too sophisticated for some students, particularly the arguments that show that, in certain circumstances, there *has* to be a particular number of flights (see Level 1, Step 2 and Level 2, Step 2). But most students should be able to understand that if the maximum number of aliens cross in the shuttle to Helios, and only one comes back, then the number of flights can't be reduced.

Table 1.11: Australian Curriculum content descriptions for the *Arks and Tarks* activity

Activity level	Problem	Content descriptions
1	Six Tarks	*Foundation* Establish understanding of the language and processes of counting by naming numbers in sequences, initially to and from 20, moving from any starting point (ACMNA001) Sort and classify familiar objects and explain the basis for these classifications. Copy, continue and create patterns with objects and drawings (ACMNA005) *Year 1* Investigate and describe number patterns formed by skip counting and patterns with objects (ACMNA018)
2	Three Arks and three Tarks	*Foundation* ACMNA001 (see above) ACMNA005 (see above) *Year 1* ACMNA018 (see above) *Year 3* Describe, continue, and create number patterns resulting from performing addition or subtraction (ACMNA060)
3	An inhabitable moon	*Year 1* ACMNA018 (see above) *Year 3* ACMNA060 (see above)

The *Arks and Tarks* problem belongs to a puzzle genre often called 'Crossing the river problems', such as the 'Farmer and fox' problem. There are many variations on this theme that could be followed up at some other time.

Note: All the questions asked in the text are directed at the students.

Big ideas

The big ideas in this activity are those of the Proficiency strands: Understanding, Fluency, Problem Solving and Reasoning. Wherever we can, we encourage the students to justify their results (Reasoning). There is also an emphasis in this activity on helping students to look for patterns.

Suggested resources

» Cards
» Pens
» Coloured stones

Problem aims

» Understand a word problem
» Think logically
» Work systematically
» Develop a notation
» Develop a justification
» Extend and generalise

Key concepts

» Understand the relation between various parts of a problem in order to solve it.
» Record and justify the answer using appropriate notation.
» Write down all possibilities and check which are possible at each stage.

Possible heuristics/strategies

» Act it out
» Trial and error
» Trial and improve
» Be systematic
» Make a table
» Exhaustive search (try all possible cases)

Special note

Least possible: a situation that cannot be improved. There are times when we want to make sure we cannot get a smaller result. In that case we are seeking the least possible solution.

Level 1: Six Tarks

Problem

Tarks are creatures from another solar system. Six Tarks are on the planet Eos. They want to get to the planet Helios by a space shuttle. Here are some things to bear in mind.

- Every space trip has to have one pilot.
- Every Tark is a qualified pilot.
- The shuttle can carry no more than four Tarks.

Is it possible to get everyone safely across to Helios?

Problem steps

Step 1

Part of the fun of this problem is the preparation. Get all the students to make a Tark costume so that they can become a Tark. Students will also enjoy constructing the shuttle out of art materials. As an alternative, the students might draw and colour in a picture of the Tarks trying to cross to Helios. All of these activities could be done in Art time.

Once the preparations are complete, talk about the conditions of the problem and let the students ask any questions. Allocate students into groups, then let each group go away to see if they can solve the problem. Give each group a chance to dress up and show that the transfer of creatures is possible. If students are unable to dress up, encourage them to use objects within the classroom or from home to be the Arks and Tarks.

With a 4-seater shuttle (or '4-shuttle') carrying six Tarks, the trip to Helios can be done in three moves (see Table 1.12).

Table 1.12: Getting all the Tarks (T) across to Helios in a 4-shuttle

	Eos	4-shuttle	Helios
At start	6T		
Trip 1 (over)		4T⟶	
After first flight	2T		4T
Trip 2 (back)		⟵1T	
After second flight	3T		3T
Trip 3 (over)		3T⟶	
After third flight			6T

This isn't the only way to do this, so ask the rest of the class to check all suggested solutions.

Note that students may have many ways to record the flights, and don't have to do it the way it is shown in Table 1.12.

Step 2

Is three the smallest number of trips that need to be made to transfer the aliens to Helios?

The answer is 'yes', and this can be confirmed in at least two ways. One way is to show that the number of voyages has to be odd. An even number of trips will see the shuttle back at Eos—and a single trip would require a 6-seater shuttle, which we don't have. The next smallest odd number is 3 and we can do it in three moves. So the least possible number of trips is therefore three. (Make sure students understand the term 'least possible' correctly.)

A second way to do this, and one that's more accessible for students, is to notice that in the first trip in Table 1.12 we take the maximum number of creatures across. In the second trip we take the fewest number back. So far we can't do better than this. In the final trip we take all of the remaining aliens across. At no stage can we do any better than we have done. So the least possible answer here is three.

Ask the students to come up with an argument for themselves. Let them discuss their arguments (this demonstrates use of Reasoning), then explain to them the methods above. Make sure that most of them understand one of those methods so that they can apply it later at this level and in the next levels. Can they describe the method in their own words?

Step 3

Suppose that the Tarks now only have access to a 3-creature shuttle. How many trips would it take to transfer everyone from Eos to Helios? We show in Table 1.13 that it can be done in five trips.

Table 1.13: Getting all the Tarks (T) across to Helios in a 3-shuttle

	Eos	3-shuttle	Helios
At start	6T		
Trip 1 (over)		3T→	
After first flight	3T		3T
Trip 2 (back)		←1T	
After second flight	4T		2T
Trip 3 (over)		3T→	
After third flight	1T		5T
Trip 4 (back)		←1T	
After fourth flight	2T		4T
Trip 5 (over)		2T→	
After fifth flight			6T

Is it possible to do the transfer in other ways? Yes. One way is to start off with 2T. Another is to replace 1T in trip 2 with 2T and make other changes later.

Did any of your students find these or any other way? Get them to show the class their solutions so that everyone can be sure that their method is correct.

Step 4

From Table 1.13 we know that five trips are possible. But can we reduce this to less than five trips?

Using either of the methods in Step 2, we can show that five is the least possible number of trips. The first method shows that we have to have an odd number of trips. It is clear that we can't get all six Tarks over in three flights, so five is the smallest possible number and we know how to do it in five flights from Table 1.13. But we also move the largest possible number of Tarks across to Helios in the first two trips and the smallest number back. This again shows that we need five flights.

Step 5

What happens if the Tarks now only have a 2-shuttle at their disposal? What is the least possible number of moves they can make to get everyone over to Helios?

In Table 1.14 we show how to do it in nine moves.

Table 1.14: Getting all the Tarks across to Helios in a 2-shuttle

	Eos	2-shuttle	Helios
At start	6T		
Trip 1 (over)		2T →	
After first flight	4T		2T
Trip 2 (back)		← 1T	
After second flight	5T		1T
Trip 3 (over)		2T →	
After third flight	3T		3T
Trip 4 (back)		← 1T	
After fourth flight	4T		2T
Trip 5 (over)		2T →	
After fifth flight	2T		4T
Trip 6 (back)		← 1T	
After sixth flight	3T		3T
Trip 7 (over)		2T →	
After seventh flight	1T		5T
Trip 8 (back)		← 1T	
After eighth flight	2T		4T
Trip 9 (over)		2T →	
After ninth flight			6T

Encourage your students to justify that this is the least possible number of trips for a 2-shuttle with six Tarks to be transferred.

Where to from here?

- Ask students what parts of the six Tarks problem did they have trouble understanding. What helped them to understand the parts they struggled with?
- Ask students which of their friends had the best way to solve this problem. How did they solve it?
- How many moves do your students think it would take to get 100 Tarks across using a 4-, 3- or 2-shuttle?
- Get the class to suggest some problems that could be considered next; for example, they might introduce more creatures.

Level 2: 3 Arks and 3 Tarks

Problem

Arks and Tarks are creatures from another solar system. Three Arks and three Tarks are on the planet Eos. They want to get to the planet Helios by a space shuttle. Some things to bear in mind are listed below.

- Every space trip has to have one pilot.
- Every creature is a qualified pilot.
- Every pilot has to leave a shuttle when it gets to a planet.
- The shuttle can carry no more than four creatures.
- At no time can there be more Arks than Tarks on a planet, otherwise the Arks will eat a Tark.

Make costumes for the aliens and a model for the shuttle.

Is it possible to get everyone safely across to Helios?

Problem steps

Step 1

As with the Level 1 activity, let students prepare by making costumes, constructing shuttles and/or drawing and colouring Arks and Tarks. When preparations are complete, talk about the conditions of the problem and let the students ask any questions. Allocate students into groups, then let each group dress up, play and try to solve the problem. Encourage students to write down their answers once they think they know how to do the crossing successfully.

With a shuttle carrying four creatures the move to Helios can be done in three moves (see Table 1.15), making sure that the Arks never outnumber the Tarks.

Table 1.15: Getting all the six aliens across to Helios in a 4-shuttle

	Eos	4-shuttle	Helios
At start	3A3T		
Trip 1 (over)		1A3T →	
After first flight	2A		1A3T
Trip 2 (back)		← 1A	
After second flight	3A		3T
Trip 3 (over)		3A →	
After third flight			3A3T

This isn't the only way to do this, but if some of your students find another method, it's possible that a Tark may get eaten along the way. So get the rest of the class to check all suggested solutions.

Once you have acted the problem out in front of the whole class, you might want the students to work in small groups using coloured stones. Whichever way you set up the problem, encourage your class to find an efficient way to record all of the flights.

Step 2

Is three the least possible number of trips that need to be made to transfer the aliens successfully to Helios? Get the students to come up with an argument for themselves.

The answer is 'yes' and this can be confirmed in at least four ways. One way is to think what would happen if there were six Tarks that had to get to Helios. This was done in Level 1, which proved that only three trips are needed without the worry of anyone being eaten—so the eating problem can't make the transfer any easier. Now we know that we can transfer three Arks and three Tarks in three trips (without any creature getting eaten—see Table 1.15) so this has to be the least possible answer.

Second, the answer has to be odd. An even number of flights will see the shuttle back at Eos, and a single trip would require a 6-shuttle, which we don't have. So the next smallest number is three, and we can do it in three moves. The least possible number of trips is therefore three.

The third, and perhaps simplest method, is to notice that in the first trip in Table 1.15 we take the maximum number of creatures across. In the second trip we take the fewest number back. So far we can't do better than this. In the final trip we take all of the remaining aliens across. At no stage can we do any better than we have done. So the least possible answer here is three.

A fourth way is to look at all the possibilities at each step. This way we can show that three trips is least possible.

Step 3

Suppose that the creatures only had access to a 3-creature shuttle. How many trips would it take to transfer everyone from Eos to Helios? We show in Table 1.16 that it can be done in five trips.

Table 1.16: Getting all the aliens across to Helios in a 3-shuttle

	Eos	3-shuttle	Helios
At start	3A3T		
Trip 1 (over)		3A →	
After first flight	3T		3A
Trip 2 (back)		← 1A	
After second flight	1A3T		2A
Trip 3 (over)		3T →	
After third flight	1A		2A3T
Trip 4 (back)		← 1T	
After fourth flight	1AT		2A2T
Trip 5 (over)		1A1T →	
After fifth flight			3A3T

Is it possible to do the transfer in other ways? Yes. One way is to start off with 2A. Another is to replace 1A in Trip 2 with 2A and make other changes later. Did any of your students find these ways or any other way? Get them to show the class their solutions so that everyone can be sure that their method is correct (and avoids some unnecessary eating).

Step 4

From Table 1.16 we know that five trips are possible. Is it possible to reduce this number so that the transfer can be done in fewer than five trips? Encourage your students to produce an argument to support this. One of the arguments explained in Step 2 should work here.

Step 5

Can we still get a successful Arks and Tarks transfer if they only have a 2-shuttle at their disposal? See Table 1.17.

Table 1.17: Getting all the aliens across to Helios in a 2-shuttle

	Eos	2-shuttle	Helios
At start	3A3T		
Trip 1 (over)		1A1T →	
After first flight	2A2T		1A1T
Trip 2 (back)		← 1T	
After second flight	2A3T		1A
Trip 3 (over)		2A →	
After third flight	3T		3A
Trip 4 (back)		← 1A	
After fourth flight	1A3T		2A
Trip 5 (over)		2T →	
After fifth flight	1A1T		2A2T
Trip 6 (back)		← 1A1T	
After sixth flight	2A2T		1A1T
Trip 7 (over)		2T →	
After seventh flight	2A		1A3T
Trip 8 (back)		← 1A	
After eighth flight	3A		3T
Trip 9 (over)		2A →	
After ninth flight	1A		2A3T
Trip 10 (back)		← 1A	
After tenth flight	2A		1A3T
Trip 11 (over)		2A →	
After eleventh flight			3A3T

Is 11 the least possible number of moves? If it is, showing it might be harder than in any of the other cases here or in Level 1. The difficulty is showing that there has to be a trip back to Eos that requires *two* creatures to be in the shuttle. This is why we can't use any of the previous arguments.

Where to from here?

- Ask the students what parts of the three Arks and three Tarks problem they had trouble understanding. What helped them to understand them?
- Ask the students which of the class do they think had the best way to solve this problem. How did they solve it?
- What is the least possible number of ways to get 10 Arks and 10 Tarks over to Helios? What about 20 or 30? There are problems here if a 2-shuttle is being used.
- What problems can the students make up that are similar to the Arks and Tarks problem?

Level 3: An inhabitable moon

Problem

There are four Arks and four Tarks on Eos and they want to go to Helios, with the usual restrictions that there cannot be more Arks than Tarks on any planet. There is only a 2-shuttle available to them. However, there is an inhabitable moon, Selene, circling Helios. Can the creatures be transferred successfully by using Selene?

Problem steps

Step 1

This may require a certain amount of experimentation, but it can be done. One method is to get three Arks to Selene by repeatedly taking two Arks across and one Ark back. This eventually leaves one Ark and three Tarks on Eos. These can all be transferred to Helios in four round trips. The Ark on Helios can then be used to pilot the shuttle to bring all the Arks home to Helios from Selene. This takes 17 trips: six to and from Selene to deposit the three Arks, five to and from Eos to Helios to move one Ark and three Tarks, and six trips to Selene from Helios and back again.

Another way to transfer the creatures is to take one Ark to Selene and leave him there (two trips); three Arks and four Tarks to Helios (13 trips); and collect the lonely Ark on Selene (two trips).

Step 2

Try as we might, we can't find a solution to get all the aliens to Helios in fewer than 17 trips. Can your class do better?

The trick here for the students is to notice that you first have to get at least one Ark to Selene, otherwise there are going to be problems. This means that moving one Ark anywhere requires two trips, whether you are going via the moon or not.

Step 3

Ask the students: Can any number of Arks and Tarks be taken across to Helios using Selene? Try five Arks and five Tarks, 10 Arks and 10 Tarks, even 100 Arks and 100 Tarks. Is there some number of creatures that won't go across, such as in the case of four Arks and four Tarks with a 2-shuttle and no moon? In that case, show it can't be done. If they can be moved across, what is the fewest number of trips that are needed?

All of these combinations should be solvable. The trick is to get one Ark to Selene first and then work on four Arks and five Tarks, nine Arks and 10 Tarks, 99 Arks and 100 Tarks and so on.

Step 4

How much more efficient is it to have a 3-shuttle or even a 4-shuttle?

Where to from here?

- There are any number of scenarios that can be explored now. What does your class suggest? Follow these up.

CHAPTER 5:
THE FARMYARD PROBLEM

Initial problem

There are only goats and chickens in the farmyard. The farmer has counted 8 animals and 26 legs. How many goats and chickens are there?

Background information

The farmyard problem is one you might already know. Given the numbers of legs and the numbers of animals in the farmyard, we have to find how many goats and chickens there are.

We are not sure of the history of this problem but it has been used in many countries over a long period. The goats and chickens are sometimes replaced by spiders and ants, bicycles and tricycles, and all kinds of other things. The problem has also been used by mathematics educators to analyse students' problem-solving abilities. You might like to ask your students to do some research online to see what interesting things they can find out about this problem.

The levels develop by going from questions where the numbers of animals and legs are given (Level 1), to exploring all of the possibilities when there are eight animals in the farmyard (Level 2), to looking at what happens if the number of goats is a multiple of the number of chickens (Level 3).

There are a lot of approaches to Level 1 and some of these can be used by Foundation students. You may want them to use farmyard animal models or plasticine animals. You may decide that initially the numbers of legs and animals could be less for Foundation and some Year 1 students.

The problems of Level 2 can be done by Year 1 students with a little help from you. Level 3 can be done by Year 3 students and may be solvable by some Year 2 students too.

Table 1.18: Australian Curriculum content descriptions for *The Farmyard Problem*

Activity level	Problem	Content descriptions
1	With the lot	*Foundation* Sort and classify familiar objects and explain the basis for these classifications. Copy, continue and create patterns with objects and drawings (ACMNA005) *Year 1* Investigate and describe number patterns formed by skip counting and patterns with objects (ACMNA018) *Year 2* Solve simple addition and subtraction problems using a range of efficient mental and written strategies (ACMNA030)
2	How many legs?	*Year 2* Describe patterns with numbers and identify missing elements (ACMNA035) *Year 3* Describe, continue and create number patterns resulting from performing addition or subtraction (ACMNA060)
3	Doubling up	*Year 2* Recognise and represent multiplication as repeated addition, groups and arrays (ACMNA031) *Year 3* Recall addition facts for single-digit numbers and related subtraction facts to develop increasingly efficient mental strategies for computation (ACMNA055) *Year 4* Develop efficient mental and written strategies and use appropriate digital technologies for multiplication and for division where there is no remainder (ACMNA076)

Big ideas

Again the big ideas in this activity are those of the Proficiency strands: Understanding, Fluency, Problem Solving and Reasoning. Wherever possible, encourage students to justify their results; this demonstrates their use of reasoning.

Four useful problem-solving tools that are used here are: acting it out, using materials (stones), guess and check and drawing a picture.

Suggested resources

- » Rope or string
- » Cards/paper
- » Coloured pencils
- » Coloured stones
- » Farmyard animal models or plasticine animals

Problem aims

- » To understand a word problem
- » To think logically
- » To think about and explore different possible solutions
- » To extend and generalise

Key concepts

- » Describe patterns with numbers
- » Create algebraic expressions
- » Understanding the relation between various parts of a problem in order to solve it
- » Using the link between different animals and the number of legs they have

Possible heuristics/strategies

- » Act it out
- » Guess and check
- » Guess and improve
- » Draw a picture
- » Make a table

Level 1: With the lot

There are only goats and chickens in the farmyard. The farmer has counted 8 animals and 26 legs. How many goats and chickens are there?

Problem steps

Step 1

Divide the class into two groups representing goats and chickens. They should be distinguishable in some way. For example, the 'goats' could kneel down on all fours while the 'chickens' stand up on two legs. Alternatively, cards could be used to identify each student as either a goat or a chicken. Part of the classroom should then represent the farmyard.

Allow students to *guess and check* the numbers of each animal needed as they *act the problem out*. This can be done by putting different numbers of goats and chickens into the farmyard so that there are eight animals altogether. Then the number of legs is checked to see whether the right number of goats and chickens has been found.

If students don't find the right answer the first time, ask them what they could do next; start them thinking about good/efficient ways to proceed. Let the children discover for themselves that if you increase the number of goats by one and decrease the number of chickens by one then the number of legs goes up by two, and vice versa. This means that students can take a guess and, if it doesn't work, they can adjust the different animals until the correct number of legs is found.

One useful alternative strategy for students is to draw circles to represent the animals. They can add two legs to each animal to start with, then they can add two legs per body to make goats until all the legs have been used up.

There are many ways to approach this question but the answer is five goats and three chickens ($5 \times 4 + 3 \times 2 = 26$). It's good practice for the students to check their answers. How can they do this? How did we just do it?

Step 2

Suppose the farmer has counted 16 animals and 50 legs. How many goats and chickens are there?

This time there are nine goats and seven chickens. Encourage the students to check their answer by using a different method to the one they used the first time.

Where to from here?

Use the following questions as part of a small group discussion. Groups can then report back to the whole class.

- How many ways did your class find to solve this problem? Which way was the best?
- Can they use that method to solve a problem where the farmer has counted 22 animals and 54 legs?
- What other problems can they invent that are like the farmyard problems that they've looked at so far?
- Do they know how to solve these problems?

Level 2: How many legs?

Problem

There are only goats and chickens in the farmyard. The farmer has counted 8 animals. How many different numbers of legs can there be?

Problem steps

Step 1

This activity asks students to find out all that they can about one scenario. What can we say if we restrict ourselves to just 8 animals and as many legs as we like? If we had exactly 8 animals, how many different problems could we make up? In this way the activity becomes a mini-investigation.

The kinaesthetic experience of moving students around in Level 1 has now probably served its purpose. For this activity it's better to use something like tokens or cards to represent the goats and chickens.

There are nine different solutions using eight animals and these are shown in Table 1.19.

Table 1.19: The different ways to get eight animals

Legs	16	18	20	22	24	26	28	30	32
Goats	0	1	2	3	4	5	6	7	8
Chickens	8	7	6	5	4	3	2	1	0

Step 2

Spend some time exploring this 'collection'. Discuss with the class what they have found out so far. What did they notice? You might expect at least the following four things:

- The total number of legs always has to be an even number.
- If you replace a chicken with a goat the number of legs increases by two.
- The maximum number of legs is 32 (8×4) and the minimum is 16 (8×2).
- With eight animals you can only create nine problems.

How many problems can you create with 7, 9, 10 or 20 animals?

These observations provide a holistic overview of the problem, which sets the scene for opening up and extending the problem further.

Step 3

If the farmer has counted 30 legs, what are the largest and smallest numbers of animals that there could be in the farmyard?

The easiest way to solve this is to note that you'll have the most animals if they're all chickens, which have two legs—in that case we'll have 15 animals. The fewest animals will be needed if we have the most goats possible. However, 4 doesn't go exactly into 30. The best we can do is to have seven goats (and one chicken). So the smallest number of animals is eight.

Step 4

Can students find a quick way to find the largest and smallest numbers of legs if the farmer has counted 13 animals?

The largest number of legs occurs if we have all goats, so $13 \times 4 = 52$ legs. The smallest number of legs happens if we have all chickens, so $13 \times 2 = 26$ legs.

Where to from here?

- What ways did your students use to find the answers for all the situations where there were eight animals? What was the 'nicest' method they used to find an answer to these problems? What does 'nice' mean here?
- What other problems can they invent that are like the farmyard problems that they've looked at so far? Do they know how to solve these problems?
- What ideas do your students have for more problems like this? How about introducing a third type of creature with a different number of legs—lobsters perhaps!

Level 3: Doubling up

Problem

This time the farmer has counted nine animals. While doing this he saw that there were twice as many chickens as there were goats. How many legs were there?

Problem steps

Step 1

This can again be solved by *acting it out, using concrete materials, drawing* and so on.

Whatever the method, two chickens can be paired with a goat. Three groups of these three animals can be put together to make nine animals. Each group has $2 \times 2 + 4 = 8$ legs. So the nine animals have $3 \times 8 = 24$ legs altogether.

Step 2

Suppose the farmer counted 20 animals, and saw that there were three times as many goats as there were chickens. How many legs would there be?

There are groups of four animals that have three goats and one chicken. Each group has $4 \times 3 + 2 \times 1 = 14$ legs. Five groups of four make 20 animals, so there are $5 \times 14 = 70$ legs.

Step 3

Now the farmer counts 200 legs. How many animals are there if the number of chickens is a multiple of the number of goats?

There are at least seven possible answers for this, so let students work for a while in small groups to see what they come up with. Whenever a group produces an answer they should write it on the board for everyone else to check. But how do they know when they have them all?

Here's one way to tackle the problem. Table 1.20 lists the possible multiples of goats, the sizes of groups that have one goat in them, the number of legs in a group, the number of goats and chickens this implies, and the total number of legs. The last three columns are used as a check; we don't actually need to go past the 'Number of groups' column. If this is a factor of 200 then we have an answer. Once we get the number of groups, the number of goats equals the number of groups and the number of chickens is the number of goats times the multiple from the first column.

Table 1.20: Some of the different ways to get 200 legs

Multiple	Group size	Legs in group	Number of groups	Number of goats	Number of chickens	Number of legs
1	2	6				
2	3	8				
3	4	10	20	20	60	200
4	5	12				
5	6	14				
6	7	16				
7	8	18				
8	9	20	10	10	80	200
9	10	22				
10	11	24				

From Table 1.20 we can see that the number of legs in a group has to go into 200 exactly. So we can use this information to save time looking at all of the possible multiples. This shows that the *smallest* possible number of legs in a group can only be 1, 2, 4, 5, 10, 20, 25, 50, 100 and 200.

- Number of legs in a group is 1: not possible.
- Number of legs in a group is 2: this would mean that the multiple was zero and we would only have chickens. Is this a reasonable answer? If it is, then there are no goats and 100 chickens. Are your students happy with this answer?
- Number of legs in a group is 4: this would mean that we would only have goats. Is this a reasonable answer? If it is, then there are 25 goats and no chickens. But what is the multiple here? Are your students happy with this answer?
- Number of legs in a group is 5: not possible.
- Number of legs in a group is 10: goats = 20 and chickens = 60
- Number of legs in a group is 20: goats = 10 and chickens = 80
- Number of legs in a group is 25: not possible.
- Number of legs in a group is 50: goats = 4 and chickens = 92
- Number of legs in a group is 100: goats = 2 and chickens = 96
- Number of legs in a group is 200: goats = 1 and chickens = 98 or goats = 0

You need to discuss with the class whether or not 'no chickens' and 'no goats' are reasonable answers given the question.

Where to from here?

- Where did your students use multiplication or division?
- What problems did they use *acting it out* to help with? Could they have done the problem without acting it out?
- What problems did they use a table to help with? Could they have done the problem without a table? Were there any problems that they could only do using a table?
- What other questions can your students invent about the farmyard? The legs in Step 3 could be changed to animals, but be careful to make sure that the number of animals has a lot of factors.

CHAPTER 6: THE 12 GAME

Initial problem

This game is played by two players with a total of 12 stones. Filipe and Sally play alternately by taking one or two stones from the table. The player who takes the last one or two stones from the table wins. If Filipe plays first, is Sally certain to win?

Always assume that both players are playing to win.

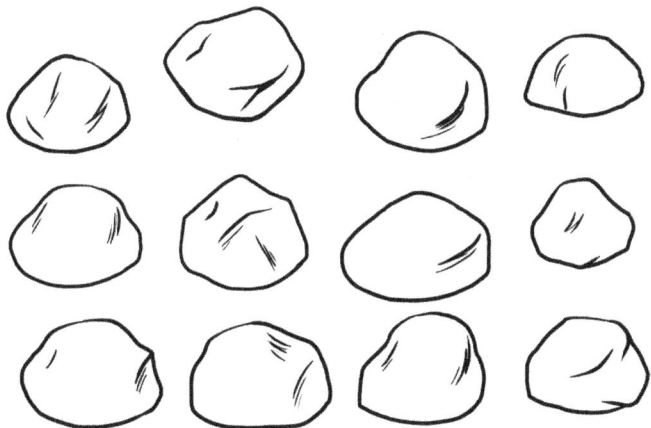

Background information

This game is from the mathematical topic called *game theory*. This topic is of particular importance in economics and other areas where analysing strategies is of great value. *The 12 game* also has strong connections to the Number and Algebra strand of the Australian Curriculum: Mathematics.

The game we use here involves 12 tokens ('stones') and two players. Each player in turn removes one or two stones. There are two variations of this game: the one who removes the last stone from the table may be the winner or the loser. Obviously the rules for who wins or loses have to be determined ahead of time.

At Level 1, the person who takes the last stone is the winner. Can the students find a strategy that enables one person to be sure of winning? The Level 1 game can be played by Foundation and Year 1 students but they will probably find it hard to articulate the pattern involved. It could be worth reducing the number of stones for them to three at first, then raising the number as they become more familiar with the game. If any of your Year 2 students find the 12 stones version hard to understand, reduce the number.

»

At Level 2, the person who takes the last stone loses. Is there a strategy? Level 2 will be fine for students in Year 2 and above who have mastered Level 1. The activity is appropriate for Year 3 students, who probably won't need any additional assistance.

At Level 3, players can remove one, two or three stones. What differences does this make to the Level 1 and Level 2 games? This level is mainly for Year 3 and Year 4 students.

The important thing to note in this activity is that the players know how to win and are trying their best to win. This forces a particular winner each time.

Table 1.21: Australian Curriculum content descriptions for *The 12 game*

Activity level	Problem	Curriculum content descriptions
1	The basic game	*Foundation* Establish understanding of the language and processes of counting by naming numbers in sequences, initially to and from 20, moving from any starting point (ACMNA001) Sort and classify familiar objects and explain the basis for these classifications. Copy, continue and create patterns with objects and drawings (ACMNA005) *Year 1* Investigate and describe number patterns formed by skip counting and patterns with objects (ACMNA018)
2	The *misère* version	*Foundation* ACMNA005 (see above) *Year 1* ACMNA018 (see above) *Year 3* Describe, continue, and create number patterns resulting from performing addition or subtraction (ACMNA060)
3	Taking more	*Year 1* ACMNA018 (see above) *Year 3* ACMNA060 (see above)

Another common, similar problem is to start counting up from 0 using the next number or the next two numbers. Again, the first person to reach 12 may be either the winner or the loser, depending on what was decided beforehand. The underlying mathematics of these two versions is the same. We gradually use all of these four variations over the three levels.

The 12 game here is also closely related to the traditional game of Nim, which has many variations and may be of Chinese origin.

Note: All the questions asked in the text are directed at the students.

Big ideas

» Seeing patterns in games

» Justifying patterns

Suggested resources

» Coloured stones

Problem aims

» To understand a word problem

» To understand about strategies for games

Key concepts

» Understanding the relation between various parts of a problem in order to solve it

» Recording and justifying the answer using appropriate notation

» The multiples of three

» The remainders on dividing by three

» Understanding that not all games need luck to win: for some games there are strategies that enable a player to force a win

Possible heuristics/strategies

» Trial and error

» Trial and improve

» Make a table

» Try small cases/special cases

» Have you seen a problem like this before?

Special note

The *misère* version of a game is one that has an opposing goal to the original game. If a game has the goal of getting the most of something, then the *misère* version will be won by finding the least of the same something.

There are some technical phrases that we use here. If you think that your students will find them difficult, please don't use them. We include them here for use with more able students.

If and only if: A result that says A is true if and only if B is true, means that if A is true then B is true and if B is true, then A is true. So there is a double implication here.

Without loss of generality: By making a given assumption, we are still tackling every case. The given assumption doesn't reduce the generality of the discussion.

Level 1: The basic game

Problem

This game is played by two players with a total of 12 stones. Filipe and Sally play alternately by taking one or two stones from the table. The player who takes the last one or two stones from the table wins. If Filipe plays first, is Sally certain to win?

Always assume that both players are playing to win.

Problem outline

Step 1

The difficult part about this game is that students may take a long time to realise that there is a winning strategy and that by playing a certain way, either the first (Filipe) or second player (Sally) must always win. No matter how many stones there are, one player will always have a winning strategy.

Let the class play in pairs for a while and record who wins each game. When they have played 10 or so games, get the whole class together and ask them to vote on the following three options.

1. The first player will always win.
2. The second player will always win.
3. It's just a matter of luck and anyone can win.

Whichever one they choose, ask them to justify their choice. This might provoke a debate, but it is likely that most students will think that it's just a matter of luck who will win.

Step 2

To get some more evidence to decide on the three options, move the pairs into small groups. Have the groups work with a few stones to address the following questions, recording the moves that they take and the conclusions they make.

- If there is only one stone, will Filipe (the first player) always win?
- If there are only two stones, will Filipe (the first player) always win?
- If there are only three stones, will Filipe (the first player) always win?
- If there are only four stones, will Filipe (the first player) always win?

Bring the class together for a discussion of the results so far. It may be useful to have two students demonstrating the various moves for the class.

With one stone, the first player has to take it and so will always win.

With two stones, the first player will win if they take both stones but will lose if they take only one. However, Filipe is trying to win, so he'll always take both stones with his first move and win.

With three stones, things change. If Filipe takes one stone, Sally will take two. If Filipe takes two stones, Sally will take one. Whatever Filipe does, Sally wins. So three stones always means a win for the second player.

Things get more complicated with four stones. If Filipe takes one stone, then Sally takes one, Filipe takes the remaining two stones and wins. If Filipe takes one stone, then Sally takes two, Filipe takes one stone and wins again. If Filipe takes two stones, Sally will seize her chance and take two to win—but because Filipe is playing to win, he will *definitely not* take two stones on his first move. Filipe will always take one stone and thus always win.

This is an important point and a subtle one, so make sure your class understands the reasoning behind Filipe's decision.

Step 3

Put the results so far in a table on the board (see Table 1.22).

Table 1.22: The winners of games with one to six stones

Number of stones	Who will win
1	F (Filipe)
2	F (Filipe)
3	S (Sally)
4	F (Filipe)
5	
6	

Now ask them what will happen if there are five or six stones on the pile. Get each group to play these games and record their results. Bring them back together to complete Table 1.22 on the board. The number of stones removed for each move has to be justified.

What does the table suggest are the best numbers for Sally? Which are the best for Filipe?

Now repeat the Step 2 vote with the three options, where 12 stones are being used. Can the students justify each of their choices?

Step 4

Your students should now have come up with Table 1.23. Ask them to justify the two new entries.

Table 1.23: The winners of games with one to six stones (complete)

Number of stones	Who will win
1	F (Filipe)
2	F (Filipe)
3	S (Sally)
4	F (Filipe)
5	F (Filipe)
6	S (Sally)

Justifying this is not easy; there are lots of stones and lots of possible moves.

Let's go back to four stones again. Ask the class: What did Filipe do before to win? He took just one stone on his first move, because if he had taken two he would have lost. But why does that work?

If he takes one, there are then three stones on the pile—and as we know from the table, three stones means a win for the second player. But when Sally comes to her move on three stones, their roles have changed—she is now the *first* player and Filipe is the *second* player, which means Filipe will win.

Can someone now see why Filipe will win if there are five stones on the pile? Let them come up with an argument. How many stones should Filipe take from the five-pile on his first move? Why?

If he takes two stones away, the pile only has three stones on it. Now Sally comes to that pile as *first* player and will lose! So Filipe will definitely win the five-pile game.

Step 5

Now get your class to get into their groups and work through all the scenarios to complete Table 1.24.

Table 1.24: The winners of games with one to 12 stones

Number of stones	Who will win
1	F (Filipe)
2	F (Filipe)
3	S (Sally)
4	F (Filipe)
5	F (Filipe)
6	S (Sally)
7	F (Filipe)
8	F (Filipe)
9	S (Sally)
10	F (Filipe)
11	F (Filipe)
12	S (Sally)

Can your students justify the entries they put in Table 1.24?

Let's do the case of 12 stones. If Filipe removes one stone, Sally takes two and the pile is reduced to nine stones. But if Filipe takes two stones, Sally takes one and the pile is still reduced to nine stones. In the same way, Sally can make sure that next time round the pile goes down to six stones and then to three. But we know that Sally, now as second player, can win if there are three stones on the pile. So Sally can win from 12 stones.

Where to from here?

- Ask the class if Sally can win if there are 13 stones on the pile. Who wins if there are 15 stones? What patterns can your students see in this game? Can they justify all of this?
- They could have a round robin competition where, in each game, they choose a number of stones at random and who should start first, and then play.
- What questions can your class make up that produce a game similar to *The 12 game*? Who will win their game?

Level 2: The *misère* version

Problem

There are 12 stones on the table. Filipe and Sally play alternately by taking one or two stones from the table. The player who takes the last stone *loses*. If Filipe plays first, is Sally certain to win?

Always assume that both players are playing to win.

Problem steps

Step 1

Note that this is the *misère* version of the game in Level 1. The 'opposite' rule is given for the winner.

Have the class play the game to understand the new rule, then draw up a table of cases with a small number of stones. (In fact, you may want the class to repeat Steps 1 to 4 of the game in Level 1.)

Let's look at three possible sets of stones to see how they work.

- Suppose that there are three stones on the table. Filipe can win by removing two stones; Sally will have to take the last stone and lose.
- If there are four stones, Filipe has two possible moves. If he takes one, then Sally will take two and win. On the other hand, if Filipe takes two stones, then Sally will take one and win again.
- In the five-stone version, Filipe can win by taking one stone on his first move. This puts Sally in the same position as Filipe was in the four-stone version. There Sally won, so here Filipe will win.

Step 2

Put Table 1.25 up on the board and have student groups complete it with their findings.

Table 1.25: The winners of *misère* games with one to 12 stones

Number of stones	Who will win
1	S
2	F
3	F
4	S
5	F
6	F
7	S
8	F
9	F
10	S
11	F
12	F

To make it clear, show how Filipe will win the 12-game. Filipe's tactics are shown in Figure 1.6. The letters F and S represent Filipe's and Sally's moves, respectively; the up and down arrows represent Sally's two possible moves, while the horizontal arrows show Filipe's best strategy is to take two stones away on his first move. The up and down arrows then show that Sally has to reduce the pile either to eight or nine stones. Whatever Sally does on this move, Filipe can reduce the pile to seven on his next move. Subsequently, whatever Sally does, Filipe can make sure that the pile goes down by three from one of his turns to the next. This continues until he leaves one stone, which Sally has to remove and loses the game.

Figure 1.6: How Filipe wins with 12 stones

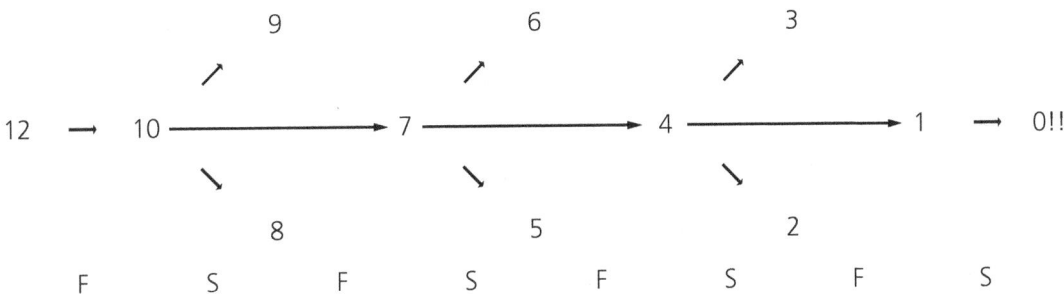

Step 3

Who does your class think will win if there are 22 stones? What about 222 stones?

Looking at Table 1.25, it seems clear that Filipe wins if the number of stones is a multiple of 3 or two more than a multiple of 3; Sally wins if the number of stones is one more than a multiple of 3.

Since 22 = 3 × 7 + 1, then Sally should win that game. Can your students confirm this by playing it out? What should Filipe's first move be?

On the other hand, 222 = 3 × 74, so Filipe should win that game. But can your class predict Filipe's best starting move?

Where to from here?

- What should Filipe's first move be to win the 11-game? What about the 111-game?
- What other patterns can your students see? Can they justify them?
- Can your students invent some games that are similar to the game played here?

Level 3: Taking more

Problem

There are 12 stones on the table. Filipe and Sally play alternately by taking 1, 2 or 3 stones from the table. The player who takes the last stone *wins*. If Filipe plays first, is he bound to win?

Always assume that both players are playing to win.

Problem steps

Step 1

The steps for the students to follow are essentially the ones in the Level 1 problem.

1. Let the students play 10 games or so to get a feel for the problem.
2. Play the game with a small number of stones.
3. Look for a pattern (conjecture what will happen).
4. See if the conjecture works for up to 12 stones by playing the games.
5. See if the students can justify the conjecture.

You can test the class's understanding for the problem by asking them to vote on the following:

- The first player will always win.
- The second player will always win.
- It's just a matter of luck and anyone can win.

Step 2

Let the class work up to Table 1.26.

Table 1.26: The winners of games with 1 to 12 stones

Number of stones	Who will win
1	F
2	F
3	F
4	S
5	F
6	F
7	F
8	S
9	F
10	F
11	F
12	S

Step 3

It looks as if Sally will win if the number of stones is a multiple of four and Filipe will win the rest. Why is this the case?

Let's look at the standard 12-stone game. Whatever Filipe does with his first move, Sally can remove one, two or three stones to make sure that the number of stones left is eight. In the same way, Sally can make sure that after her next move there are only four stones left and we know that Sally can win the four-game.

Similar, Sally follows this method for *any* multiple of four stones.

Step 4

What strategy does Filipe need to follow?

- If the number of stones is one more than a multiple of four, then Filipe takes one stone to reduce the problem to a multiple of four. From there Filipe does what Sally did in Step 3.
- If the number of stones is two more than a multiple of four, then Filipe takes two stones to reduce the problem to a multiple of four. From there Filipe does what Sally did in Step 3.
- If the number of stones is three more than a multiple of four, then Filipe takes three stones to reduce the problem to a multiple of four. From there Filipe does what Sally did in Step 3.

Where to from here?

- What patterns did your students see in this game? Do they now know why they work?
- What is the *misère* version of this game? Which numbers of stones will give Sally a win in the *misère* game?
- What games can your students make up that are like the one here?

PART 2: MEASUREMENT AND GEOMETRY

Part 2 presents three activities based on the Measurement and Geometry strand.

Table 2.1: Measurement and Geometry activities

Problem	Big Ideas
Classroom shapes	- 2D shapes - Triangles, rectangles, squares, circles - 3D objects - Pyramids, cuboids (rectangular prisms), cubes, spheres
Jenny's jelly bean problem	- Area, volume/capacity - Nets - Place value
Horrible Hal's humungous hall	- Tessellations - Symmetry - Infinity

Some reminders before you use these tasks in your classroom:

1. The questions in the text are ones you can ask your students. You're likely to be able to produce similar, more immediately relevant ones for your particular students as you work on these activities with them.
2. We have given suggested links to the Years in the Australian Curriculum: Mathematics for all the Levels in each activity. Because there will be a spread of ability in your classes, you should take these as a guide only. Take the opportunity to encourage every student to the edge of their comfort zone.
3. To take all students further, sometimes you can omit some of the later steps of a Level in favour of the early steps in the following Level.

CHAPTER 7: CLASSROOM SHAPES

Initial problem

What shapes can you see in the classroom? Can you describe them?

Background information

The three levels of this activity concentrate on 2-dimensional shapes and 3-dimensional objects (see Table 2.2).

In Level 1 we explore the places where we might see 2-dimensional objects such as squares, rectangles and circles. Level 1 is accessible for all students. More able students are likely to give deeper responses, but whatever a student says that makes sense should be praised. However, you can follow this praise by urging the student to say why they responded in that way and then asking them to make their comments more precise. Students who make incorrect statements should be treated the same way in an attempt to lead them to a better realisation of the truth. This activity may be something that you come back to from time to time and extend further.

In Level 2 we move on to objects with four, five and six straight sides. Level 2 is also available for all students. However, you would probably want to come back to this for Foundation and Year 1 students a number of times during the year. That way their knowledge and skills can be improved as their maturity increases.

In Level 3 we concentrate on solid objects such as boxes, cubes, pyramids and prisms, and note the 2-dimensional objects from Level 1 that appear as faces of the solid shapes. Level 3 gives students a chance to revise the work of the previous two levels. This level requires more maturity; parts of this activity can be done by Foundation and Year 1 students, more can be done by Year 2 students, but the bulk of it is aimed at Year 3 students.

Throughout this activity we are interested in examples of all of these objects, their properties and how they differ from other objects.

Table 2.2: Australian Curriculum content descriptions for the *Classroom shapes* activity

Activity level	Problem	Content descriptions
1	Simple flat shapes	*Foundation* Sort, describe and name familiar two-dimensional shapes and three-dimensional objects in the environment (ACMMG009) *Year 1* Recognise and classify familiar two-dimensional shapes and three-dimensional objects using obvious features (ACMMG022) *Year 2* Describe and draw two-dimensional shapes, with and without digital technologies (ACMMG042)
2	More flat shapes	*Foundation* ACMMG009 (see above) *Year 1* ACMMG022 (see above) *Year 2* ACMMG042 (see above)
3	Solid shapes	*Year 1* ACMMG022 (see above) *Year 2* Describe the features of three-dimensional objects (ACMMG043) *Year 3* Make models of three-dimensional objects and describe key features (ACMMG063) *Year 4* Compare the areas of regular and irregular shapes by informal means (ACMMG087)

The overall aim of this activity is to enable students to name, describe and become better acquainted with basic shapes and objects. We do this by looking at simple geometric objects and giving students a chance to find them, collect photos of them, play with them and discuss their similarities and differences.

The activity doesn't have to be completed in a series of consecutive lessons. Parts of it can be used from time to time as they fit into your program. You probably won't want to talk about the 2-dimensional shapes and 3-dimensional objects too close together. If nothing else, the 3-dimensional objects can be used to remind students of the 2-dimensional shapes—partly because the 2-dimensional faces help to describe the 3-dimensional objects, and partly because the nets of the 3-dimensional objects use 2-dimensional shapes from Level 1.

There is a great deal of information about simple geometric objects online, so students might be encouraged to use these sites as a resource. The Maths is Fun website has an animated description of several nets.

Big ideas

That some of the 'nicer' shapes around us have names and can be described accurately.

Suggested resources

- Paint, pens and paper
- Pictures, paper and other rectangular shapes
- Sticky tape
- Scissors
- Ruler
- Compasses (or string and pins)
- MAB cubes
- Unifix blocks
- Shoe box
- Brick

Problem aims

- Name and recognise rectangles, squares, triangles, circles, cuboids, cubes, regular shapes, pyramids, prisms and spheres.
- See where and how they are used.
- State the basic properties of these objects.
- Be familiar with and describe different basic objects.

Key concepts

- Squares
- Rectangles
- Triangles
- Quadrilaterals (4-sided figures)
- Regular shapes
- Circles
- Cuboid, box
- Cube
- Pyramid
- Prisms
- Sphere
- These terms can be described precisely

Possible heuristics/strategies

- Make a picture
- Make a model
- Exhaustive search (try all possible cases)
- Be systematic

Level 1: Simple flat shapes

Problem

What shapes can you see in the classroom? Can you describe them?

Problem steps

Step 1

Ask the class to describe a shape. Can they describe any other shapes? Which shapes are flat and which are not flat? (Introduce the idea of two and three dimensions when appropriate.) Talk about the shape of a 2-dimensional object as its outline. What flat shapes do they know the names of? Let them give examples if they can.

Ask students to name the outline of a piece of paper. What other objects can they think of that have the shape of a rectangle (oblong)?

Develop the conversation to include squares, triangles and circles. If they know the names of other shapes, include these too.

Ask students if parts of objects have flat shapes. What part of a die has a square shape?

As a class, produce a table like that of Table 2.3. For each shape the first column should have the name of an example and the second a drawing, photograph, clip art image or similar.

Table 2.3: Illustrations of some basic shapes

Rectangle		Square		Triangle		Circle	
Object	Picture	Object	Picture	Object	Picture	Object	Picture
Door		Windows		Crane's arm		Coin	
Stamp		Part of a football field				Wheels/ hubcaps	
TV						Watch face	
						Netball hoop	

Let students work in pairs to complete their own table using objects that they see in their daily lives. Encourage them to bring pictures or examples of these to school the next day. One example would be photographs of parts of buildings, which could be found online with an image search, or through a tour of your local town or city.

Step 2

Can your students see objects in the classroom with shapes other than the four that you've previously talked about? Why are they not one of the four?

Get each pair of students to list objects that don't fit into at least one of the four categories. They should also be able to say why their object doesn't fit. For example, mobile phones don't fit any category; they don't have pointy bits, nor are they round all the way round.

Step 3

Ask students how they could describe each of the four shapes. What properties do the shapes have that distinguish them from each other and from other shapes?

Discuss what all rectangles have in common. What makes two rectangles different? What makes rectangles different from any other shape? Start a discussion with the class to see how far you can go to a precise answer. The corner angles and the lengths of opposite sides are important.

Let your students work in small groups to repeat the rectangle discussion with squares, triangles and circles. Bring them back as a whole class to discuss the definitions.

Suppose they say that an object with four sides is a *rectangle*. But doesn't a square have four sides? (Yes, but this is a special rectangle with four equal sides.) Can they think of or draw a shape with four straight sides that isn't a rectangle? How does a rectangle differ from these shapes?

How could you define a triangle? Are these the only shapes that have three sides? Does that cover everything? How about a sail shape like Figure 2.1?

Figure 2.1: A non-triangular shape with three sides

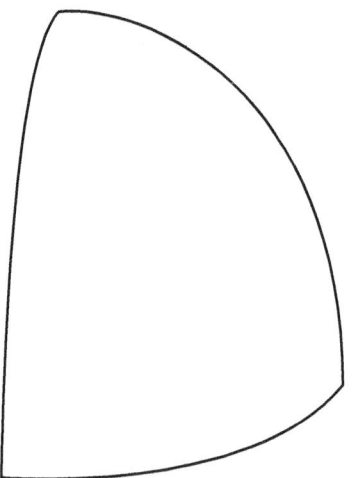

What if they suggest a shape made with *three straight sides* is a triangle? Are all of the shapes in Figure 2.2 triangles?

Figure 2.2: Are all these shapes triangles?

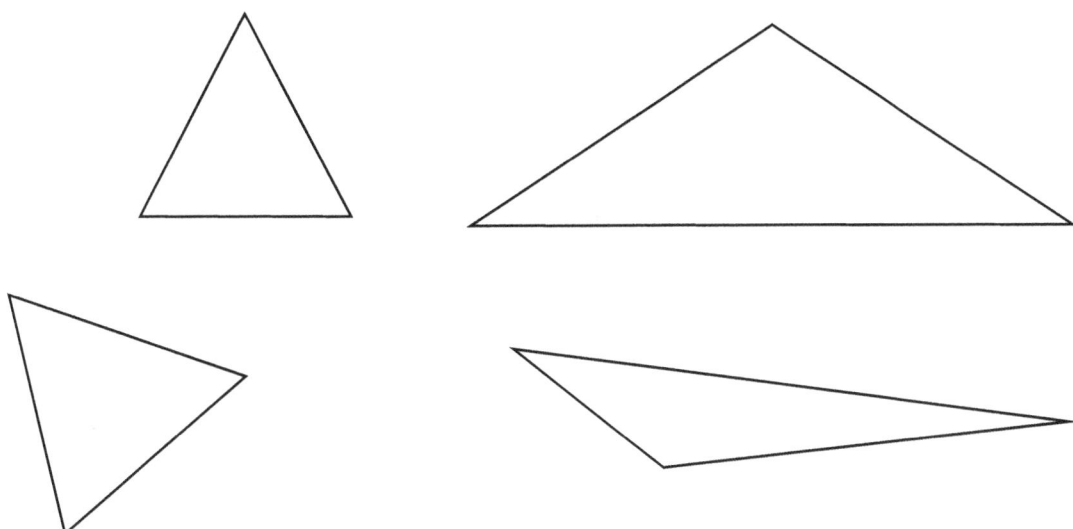

And is anything with only *one* side a circle? (Of course, this isn't a straight side.) What is the important feature of a circle? Have a look at some coins or other objects that are circles. Lead them towards the notion that circles have a central point that is the same distance from every point on the shape—but don't be in a hurry to reach this stage. Give them a chance to think about the properties a circle has. Do the same for the other shapes. Let them talk about it at home.

Step 4

Discuss how they might draw a triangle. To do this well, what devices might they have to use? They will need straight lines; a ruler will be fine. Now let them go off in groups to draw, as carefully as they can, a rectangle that isn't a square.

Repeat this exercise with squares and circles. Note that circles can be made with string with a pencil attached and a pin to tie one end of the string to a fixed spot.

Step 5

Let the students paint or draw a picture using at least three rectangles, three triangles and three circles. Do they produce abstract or real art?

Let them discuss their art.

Where to from here?

- What differences can your students think of between a square and a circle? What do these two shapes have in common? (Some degrees of symmetry.)
- What shapes other than squares and rectangles do they know that have four sides? How many can they think of? Can they name them?
- What other flat objects do they know?

Level 2: More flat shapes

Problem

How many different shapes can your students draw that have exactly *four* straight sides?

Are there any two shapes that are, in some sense, the same?

Problem steps

Step 1

Put the problem to your students and let them draw as many shapes as they can.

Discuss their results. How many did they get? Can they name them? Did they get any that aren't in Figure 2.3? Can they name any shapes that don't usually get a name? (For example, the one we say has 'no special name' in Figure 2.3 could be thought of as a 'squashed kite'.)

Figure 2.3: Some different 4-sided figures

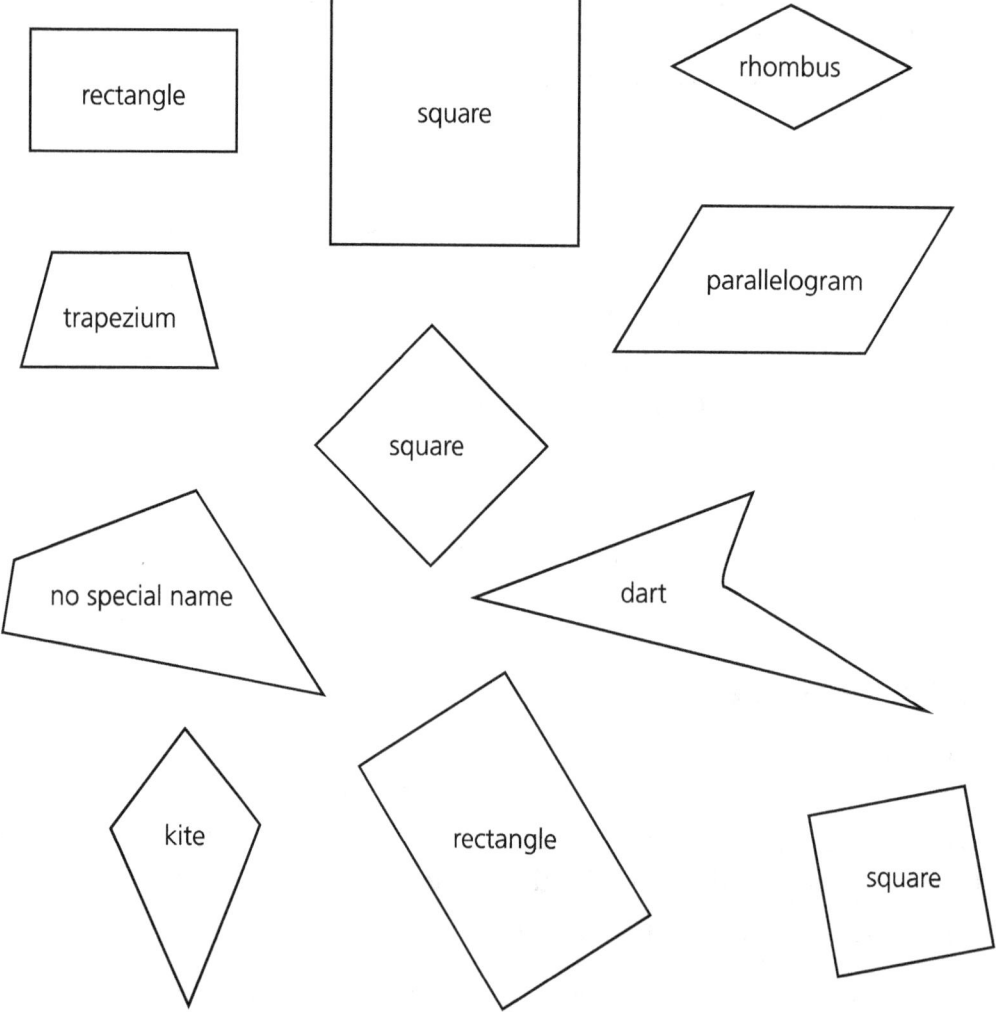

Discuss why shapes are different. How many are rectangles? What is different about rectangles? What is the significance of the corners? Are they all the same? What do they look like? Can this be incorporated into a definition of rectangle? Why are squares rectangles but rectangles are not squares? Can a triangle have the same sort of corner that rectangles have? Can other shapes have this sort of corner?

Ask questions about four-sided shapes, such as these listed below.

- Which shapes have at least one right angle/quarter turn?
- Which shapes have at least two right angles?
- Which shapes have at least three right angles?
- Which shapes have exactly four right angles?
- Which shapes have parallel lines? (What are parallel lines?)
- Which shapes have at least three parallel lines?
- Which shapes have at least two sides equal?
- Which shapes have at least three sides equal?

Step 2

Ask students what they think of this definition: a *rectangle* is any shape that has four straight sides and four corners that are the same.

Can any of your students find a counterexample to this? In other words, can anyone find a shape that has four straight sides and four corners that are the same but is *not* a rectangle?

They might be able to add that the corners have angles that are quarter turns. Does this get rid of the counter-examples? Are there no other shapes with this definition? Ask them to look up the definition of 'rectangle' online.

What is a square then? How would you define a square? Can you define a square as a particular type of rectangle? Again, this is something that can be looked up online.

You may get the following as working definitions.

Rectangle: A shape with four straight sides and four equal corners. (The corners are all quarter turns or right angles.)

Square: A rectangle with all sides the same length.

Triangle: A shape with three straight sides.

Circle: A shape, all of whose points are the same distance away from a point called the centre of the circle.

Come back to these definitions at some time to see if the students can improve them.

Step 3

Can the students make up definitions for all of the other named shapes in Figure 2.3? Discuss the suggestions that they come up with. Make sure that they understand that a definition must *only* apply to the object being defined.

Step 4

Some figures are 'nicer' than others. Why do the students think that some shapes are nice? For instance, a shape with all sides equal and all angles equal might be nicer than one that doesn't have equal sides or angles. (This is a loose idea. Let the students decide what 'nice' means.)

Shapes that are symmetric in some sense might also be nice. Get students to find as many 'nice' shapes as they can. Let them know that they are not limited by the number of sides. Discuss what they find. They will mostly likely come up with (equilateral) triangles, squares and rhombuses (kites with equal sides are rhombuses).

Step 5

Some figures are 'neater' than others. (Again, let students define what 'neat' means.) For instance, a shape with any number of sides and with all angles the same is neater than one that doesn't have equal angles. Get them to find as many 'neat' shapes as they can. Discuss.

Step 6

Some shapes, with three or more sides, are called *regular* if they are both nice and neat. Can they be more precise than that? Get the students to find as many regular shapes as they can. Discuss their answers.

Be sure that pentagons and hexagons come into this discussion. Does anyone know about regular octagons?

Where to from here?

- What flat shapes have something in common with a rhombus? (Squares have four equal sides; they also have opposite sides parallel; a rhombus is a special type of kite.)
- Pose a problem like this: Kwa has a square piece of toast every morning for breakfast. He cuts it with one straight slice. What shapes can he make this way? Can Kwa make all of the shapes we have talked about above? Why? Why not?
- Are all nice shapes neat? Are all neat shapes nice?
- Are there irregular pentagons and irregular hexagons?

Level 3: Solid shapes

Problem

What does your class know about pyramids?

Problem steps

Note:

1. We start off thinking about pyramids as the shape of tombs from ancient Egypt, which have square bases and (isosceles) triangular sides. Strictly speaking, a pyramid can have any polygon as its base; for example, pentagonal pyramids have bases that are regular pentagons.
2. We have used the term 'box' to describe rectangular prisms in this activity because it's a commonly used word and students don't have to worry about technical terms. There are, of course, shapes other than rectangular prisms that are used by manufacturers for boxes, and you may need to watch out for confusion from some students.

Step 1

Ask the class what they know about pyramids. Discuss pyramids and show them pictures. Do they know of any modern buildings that are pyramids or have parts that look like pyramids?

(You can find many images of pyramids online, both classical Egyptian tombs and modern examples such as the 'Dubai pyramid', the Louvre or even the Shrine of Remembrance.)

Step 2

Can your class tell you how many sides/faces, edges and corners a pyramid has? (The answer is 5, 8 and 5.)

Ask students to describe/name the shapes of the sides/faces of a pyramid. What is the shape of the base?

We generally think of a pyramid as having a square base. The sides/faces are triangles that have one edge in common with the base. All of the triangles have a common point, usually at the top of the pyramid. Two sides of the face triangles have the same length (i.e. they are isosceles). But do they have to have those kinds of faces?

Step 3

Ask students how they can make a pyramid. Let them discuss that. If necessary, show them a pyramid that you have made. This can be done by copying the diagram (net) in Figure 2.4, folding up the triangles to meet and then sticking the sides of the faces together.

(It may help to add tags to the sides of the triangles to make it easier to stick the sides together.)

Figure 2.4: A net of a pyramid

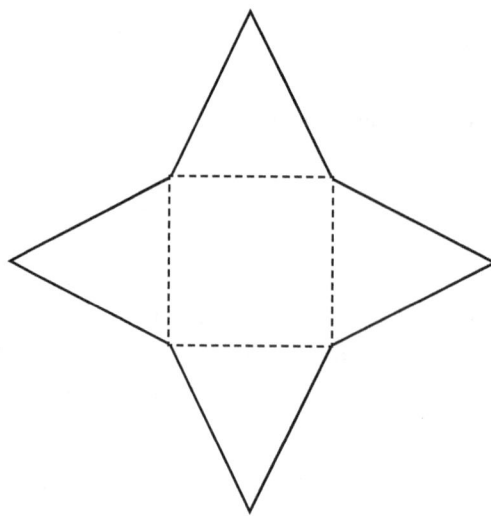

Ask a student to check that your pyramid is a pyramid, then cut along the sides of the triangular faces. Fold down the sides to make a flat shape—the net of the pyramid—so that students can make the connection between the net and the solid.

Give everyone a net and let them construct their own pyramids. Depending on the ability of your students, you might ask them to construct their own nets first.

(Students could construct their pyramid for a specific purpose. For example, they might create a pyramid from cloth-covered shapes and fill them with objects or decorations to give to their parents/guardians.)

Step 4

Can the class make a pyramid that doesn't have a square base? How would they make a pyramid with a rectangular base?

We give a net for this in Figure 2.5, but it would be good if students could build their own pyramids without using this net.

Be careful to make sure that the heights of all of the triangles are the same. Who can build the highest pyramid?

Figure 2.5: A net of a pyramid with a rectangular base

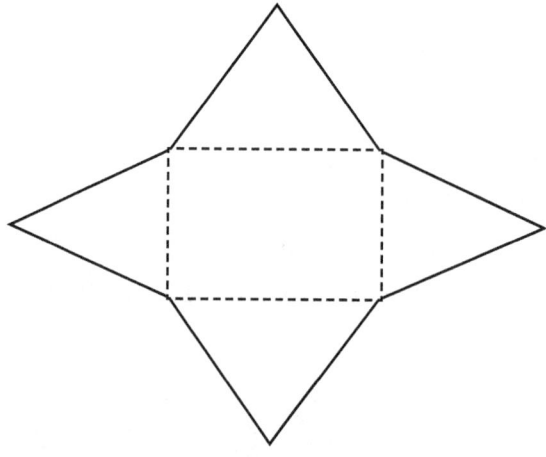

Step 5

Show the class a shoe box and a brick. Can they think of any other objects that have the same shape as these? (Boxes in general, steps, platforms, dice etc.)

Discuss this. Possibly tell them that one 'formal' name for this box shape is *cuboid* and another is a *rectangular prism.*

Step 6

- What can they tell you about these box/cuboid/rectangular prism shapes? How many faces, edges and corners do they have? (6, 12, 8)
- What shapes are the faces? (Rectangles and/or squares.)
- Are any of these faces the same? (Opposite sides will always be the same. If one face is a square then the opposite face is a square too. The other four faces will all be the same.)
- What name is given to a cuboid for which all of the faces are squares? (Cube.)

Step 7

- Can the students make a cuboid? How would they go about it? What sides have to be the same?
- We have made a net in Figure 2.6 to help with this, but they should be encouraged to make their own nets.

Figure 2.6: A net of a box

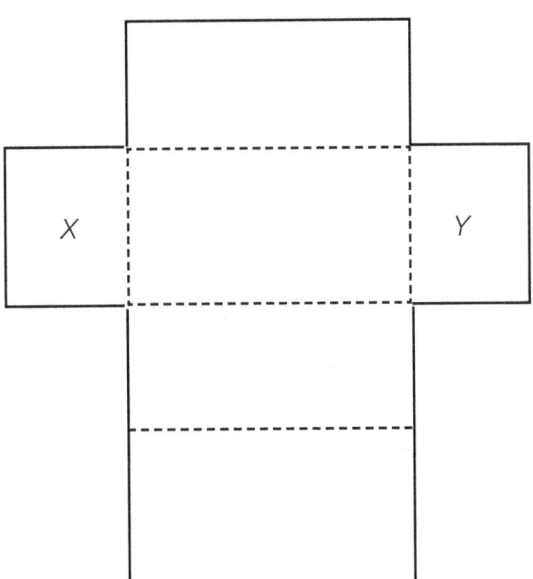

Step 8

The net of the box in Figure 2.6 has two 'flaps' to the right and the left, X and Y, that are rectangles. Suppose that we change them to pentagons. How can your class alter the rest of the net so that it is the net of another shape? Let them go away and construct this.

(They will need to add another rectangle. They also need to make sure that the lengths of the vertical sides of the 'middle' rectangles match the sides of the pentagon.)

Step 9

Tell students that shapes like this that have a straight-sided shape (rectangle, pentagon etc.) in the place of X and Y, and rectangles joining up the corners/vertices of these shapes, are called *prisms*. If the ends of the prism are rectangles we get a rectangular prism; if they are pentagons we get a pentagonal prism, and so on.

Can they produce a net and make a prism that has a regular hexagon on each end? This is called a hexagonal prism. Can they make a pentagonal prism or a square prism? What is the difference between a prism and a pyramid?

Step 10

How many different names can they find for boxes? (Rectangular prism, cuboid, some are cubes.)

So far we have only been talking about the box whose net is in Figure 2.6. Can you find shapes that are used to 'box' things in that aren't this shape? Get your students to collect (or draw) and name such 'boxes'. Note that we have been using the word 'box' loosely up to this point.

What advantages do cuboids have as boxes over other shapes? (Easier to pack together for sending out to shops.)

Where to from here?

- Ask: What properties do cuboids, cubes, pyramids and prisms have in common? How do they differ?

- Get the class to investigate the shapes they have looked at and total the number of corners plus the number of faces minus the number of edges for each one. They should always get 2. Is this the same if they combine shapes? For example, what if they put a pyramid with a square base on top of a cube, or a pyramid on top of a prism? Does it matter if corresponding faces are not the same size?

- It might be interesting to construct a really big pyramid. The class can do this by first starting with a large (A2) piece of card and cutting a piece off one end to make a square for the base. Then cut out four triangles from other pieces of card of the same size. The diagram in Figure 2.7 suggests how this might be done. Take X to be the midpoint of the shorter side. (X can be found by measuring the length of the side that it is on.) Draw lines from X to the corners A and B. Cut along the dashed lines XA and XB to get a face of the pyramid. Make three more such faces and join them to the base with sticky tape.

Figure 2.7: An aid to making a large pyramid

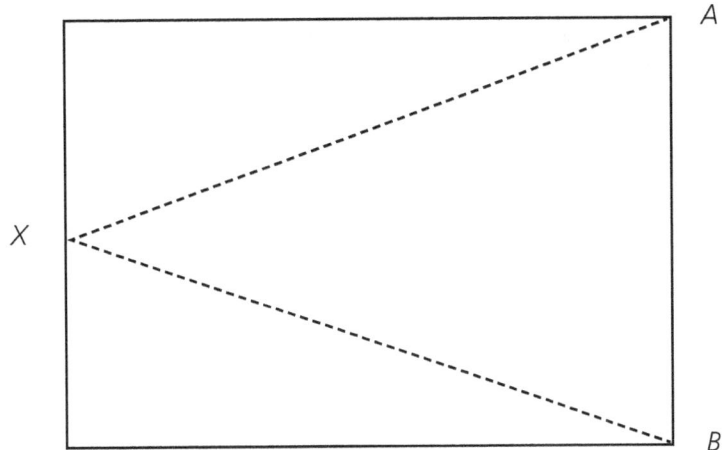

- Your class could also make a large cuboid (or several large boxes).
- Angelo likes sawing through objects. What flat shapes can he make by sawing through a solid cuboid, a solid cube, a solid pyramid or a solid prism?
- How could the class make a pentagonal pyramid or a hexagonal pyramid?
- What 3-dimensional objects have we not talked about yet? Can they get hold of photographs of them? In what way do they think these are the same as or different to the shapes we have talked about here?

CHAPTER 8:
JENNY'S JELLY BEAN PROBLEM

Initial problem

Jenny wanted to make a container that held at least 30 jelly beans. She used paper and sticky tape. How do you think she made it?

Background information

This activity is about constructing shapes that produce a given volume, specified by the number of jelly beans that a container will hold. However, we hardly use the word 'volume' at all; the emphasis is on the fixed number of jelly beans that implicitly specifies the volume. It's up to you whether you want to use or introduce the word during the activity. (Also note that we haven't specified the shape of the container.)

At Level 1, the students choose the shape of their container, though rectangular prisms and cylinders are the most likely choices as they are relatively easy to make. Level 1 is for all students. At Foundation and Year 1 the 'boxes' can be made in any way they like. The important part here is that students experience making boxes and get some feel for volume.

At Level 2 the containers are rectangular prisms, cylinders and then cones. Level 2 is similar to Level 1 but with more instructions about the shapes of the box to be used.

The shapes in Levels 1 and 2 can be constructed without the use of nets, but students from Year 2 should be able to make the boxes from nets.

At Level 3 the students make their own lidless rectangular prism by cutting square pieces out of a rectangle. The students choose the size of the squares they cut here. They're encouraged to use MAB minis to find the capacity of their box, and to see what variations in volume result by changing the size of the square that they cut out. Eventually we want them to maximise the volume. (Note that Level 3 contains a crossover to place value when they can count MAB blocks using blocks, longs and flats.) Level 3 might be attempted by younger students but it is aimed at Years 3 and 4; it requires careful measuring and is an introduction to the equation of volume for a cuboid.

Table 2.4: Australian Curriculum content descriptions for *Jenny's jelly bean problem*

Activity level	Problem	Curriculum content descriptions
1	30 jelly beans	*Foundation* Use direct and indirect comparisons to decide which is longer, heavier or holds more, and explain reasoning in everyday language (ACMMG006) Establish understanding of the language and processes of counting by naming numbers in sequences, initially to and from 20, moving from any starting point (ACMNA001) Connect number names, numerals and quantities, including zero, initially up to 10 and then beyond (ACMNA002) *Year 1* Measure and compare the lengths and capacities of pairs of objects using uniform informal units (ACMMG019) Develop confidence with number sequences to and from 100 by ones from any starting point. Skip count by twos, fives and tens starting from zero (ACMNA012)
2	50 jelly beans	*Year 2* Compare and order several shapes and objects based on length, area, volume and capacity using appropriate uniform informal units (ACMMG037) *Year 3* Measure, order and compare objects using familiar metric units of length, mass and capacity (ACMMG061)
3	The biggest box	*Year 4* Use scaled instruments to measure and compare lengths, masses, capacities and temperatures (ACMMG084) Compare objects using familiar metric units of area and volume (ACMMG290)

Jenny's problem is about creating containers to hold jelly beans. However, you can use anything that is readily available to fill the containers. At Level 3 we use MAB blocks as these are better for getting an accurate value of the size of the boxes to be constructed.

(If students have food allergies or restrictions, you might want to think about using pasta shapes rather than jelly beans.)

Also don't feel constrained by the number of jelly beans or blocks. The use of the numbers 30 and 50 in Levels 1 and 2 just represents an increase in the number of objects that might be used; use whatever numbers you judge appropriate for your class.

Big ideas
» Volume
» How different shapes have the same volume
» How similar shapes can be changed to give different volumes

Suggested resources
» Jelly beans
» MAB blocks
» Paint, pens and paper
» Sticky tape
» Scissors
» Ruler

Problem aims
» To see that different objects can contain the same volume
» To make a box without a lid not using a net
» To investigate how the volume of the lidless box varies using different sizes of squares

Key concepts
» Cuboid, box
» Cylinder
» Cone
» Square
» Rectangle
» Volume

Possible heuristics/strategies
» Make a model
» Estimate
» Be systematic

Level 1: Thirty jelly beans

Problem

Jenny wanted to make a container that held at least 30 jelly beans. She used paper and sticky tape. How do you think she made it?

Problem steps

Step 1

There are a number of ways that Jenny could do this, but a box (cuboid) shape or a cylinder shape are the likeliest solutions. However, don't prejudice students' work by insisting on a specific shape. Let them work in small groups to come up with their own ideas.

Get the groups to each make two containers that are different shapes. Ask them how many jelly beans they hold? Check that each container holds at least 30 jelly beans.

How could they change their shapes to make them hold a number of beans closer to 30? Let them have another try to get closer to holding 30 jelly beans.

Step 2

When the containers have been finished, put them on display. Check the containers to make sure that they hold at least 30 jelly beans. Discuss the containers with the students: How did you make them? How did you know that they hold at least 30 jelly beans? How many do they actually hold?

Can the students put them in order of height so that the tallest is on the left and the shortest on the right? Does the tallest one hold more than the shortest one?

Step 3

Is it possible for the students to make a short container that can hold more jelly beans than a tall one? (Yes.)

Ask the class how this can possibly be. (The short one has to be longer or wider than the tall one.)

If the students have problems with this concept, make some short and tall containers to demonstrate. How short can they make a container? Is it possible to make a very short container that would hold 1000 jelly beans? How? (It would have to be very wide.)

Where to from here?

- Which containers do the students think was the 'nicest' in some way?
- What container did they think was the nicest shape?
- If they were going to do this again, what shape would they choose for their containers?
- Can your students make a container that isn't a box or a cylinder? (How about a cone?) Is it possible to make this so that it holds *exactly* 30 jelly beans?
- What other ideas do your students have?

Level 2: Fifty jelly beans

Problem

Jenny wants to make a box that holds as close to 50 jelly beans as she can. How should she do this?

Problem steps

Step 1

Run a competition for your students. The idea is to make a box (of any shape and with no lid) that will hold as close to 50 jelly beans as possible.

After students make their boxes, ask each student to give points to each box. A very good box should get 3 points, a good box 2, and an okay box 1. There should be a bonus of 10 points if 46 to 54 jelly beans (or whatever tolerance you decide is acceptable) will go into the box without the beans going higher than the top of the box. The student with the highest score wins.

Step 2

Run a second competition, but this time use a cylindrical container.

How can students make sure that exactly the right number of jelly beans is contained in the cylinder? (One possible solution: make the container too large at first, then cut part of the open end of the cylinder off so that it contains fewer beans.)

Step 3

Run the competition for a third time, this time using a cone as the shape of the container. How can they make a cone? Let them discuss this before they start building one.

How can students make sure that exactly the right number of jelly beans is contained in the cone? (The solution from Step 2 is one possible approach.)

Step 4

Which of these shapes appears to use the least amount of paper? How can the class judge the size of the paper used? Is there more than one way to do this?

Where to from here?

- Ask the students what they discovered about adjusting containers to hold a given number of jelly beans. Will these methods work for *any* shaped container?
- What other shaped containers can the class make? Could these shapes be more accurate than boxes, cylinders or cones?
- What other ideas do your students have?

Level 3: The biggest box

Problem

Jenny is making a cuboid box. She knows that she can make one from a rectangular sheet of paper (see Figure 2.8) by cutting the shaded squares from each corner and folding along the dashed lines. Can your students make a container this way?

Figure 2.8: How to make a box from a rectangle

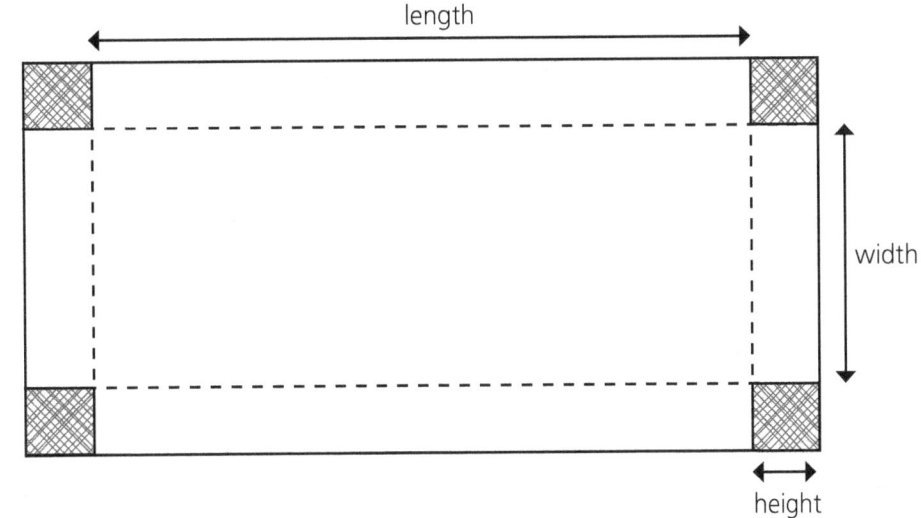

Problem steps

Step 1

Jenny measures out a square in each corner of the rectangle. She folds along the dotted lines and turns up the sides to make a box, then used sticky tape to stick the sides together. When she's done she has a box that has the length, height and width shown. The size of each dimension will vary depending on the sizes of the paper used and the squares cut out.

Using any rectangle and any sized square that they like, get the students to make a box with this method. How many MAB blocks does their box hold?

(The counting here can be simplified if they use MAB flats and longs.)

Step 2

Get the students to make another box using Jenny's method. Keep the rectangle the same as in Step 1 but cut off different-sized squares. How many blocks does this new box hold? Is it more or less than the first one? Why? How many more blocks?

This may be hard to answer exactly because students won't know that the volume of a box is the product of length, height and width. However, they can count the number of MAB cubes it holds. Again, this can be speeded up by using flats and longs. (Doing this provides an entry into place value too.) As the cut-out square gets bigger, will the number of blocks that the container can hold get bigger or smaller?

Put the results of the class in a table on the board so that everyone can see them. The table should show the length, width and height of each box, as well as the number of blocks it will hold.

The results should show that if you start with a small square and make it bigger, the number of blocks it holds will get bigger for a while, then start to get smaller. Identifying where the change occurs will enable students to find the size of the square that gives the box with the largest volume.

Step 3

Download the file of a 14 by 20 rectangle from the series website for the students to make rectangles with. Ask them to only cut the squares that are marked by dotted lines.

Get everyone to guess the size of the square that is needed to make a box that will hold 200 blocks.

Ask them to build a box to check out their conjectures. What conclusions did they come to?

Step 4

What is the largest number of blocks that your students can pack in a rectangle that uses Jenny's method on the downloaded template? (This template should only be cut along the dotted lines.)

Get them all to make a guess and record these guesses on the board. What guesses do they come up with? How would they check these guesses?

To check their guesses let them all work on one or two variations of square size, but make sure that all the squares from size 1 to 7 are used. Make up a table such as the completed one in Table 2.5 to record results. Encourage them to use flats and longs to speed up the counting.

Table 2.5: Completed table of square sizes and box dimensions

Square size	Width	Length	Number of blocks
1	12	18	216
2	10	16	320
3	8	14	336
4	6	12	288
5	4	10	200
6	2	8	96
7	0	6	0

What rectangle holds the smallest number of blocks?

Where to from here?

- What did your students discover when doing this problem?
- What ideas do your students have for making problems using Jenny's method of building boxes or some other method of building boxes?
- Your students have been seeing what happens if you change the height of their box. What happens if they change the width or length?

CHAPTER 9:
HORRIBLE HAL'S HUMUNGOUS HALL

Initial problem

Horrible Hal's humungous hall
Is as long and as wide as the sky is tall.
You can run all day, you can run all night,
But never the end will come into sight.
The walls may be red, the walls may be green,
Nobody knows as they've never been seen.

Can you describe Horrible Hal's humungous hall?

Background information

Students will have seen bricks making up walls, and tiles covering floors and walls. In *Horrible Hal's humungous* hall we look at what shapes can tile never-ending floors and ceilings.

In Level 1 we concentrate mainly on triangles. This is largely an exploration of tiling with different shapes, so it can be done by students at all levels of the curriculum. However, you should expect more mathematical discussions from more able students (at either a higher year level or in a given class).

In Level 2 we look at squares and their variations. Again, this activity is about experimenting, but the steps starting at Step 5 will be difficult for students in Foundation and Year 1.

»

In Level 3 we do a thorough investigation of all 4-sided shapes. Level 3 is more accessible to Year 3 and 4 students because of the concepts involved, but a lot of it can still be explored by students who are not yet at that level.

Table 2.6: Australian Curriculum content descriptions for *Horrible Hal's humungous hall*

Activity level	Problem	Content descriptions
1	Hal's floor	*Foundation* Sort, describe and name familiar two-dimensional shapes and three-dimensional objects in the environment (ACMMG009) *Year 2* Investigate the effect of one-step slides and flips with and without digital technologies (ACMMG045)
2	Hal's ceiling I	*Year 3* Identify symmetry in the environment (ACMMG066)
3	Hal's ceiling II	*Year 3* ACMMG066 (see above) *Year 4* Compare and describe two-dimensional shapes that result from combining and splitting common shapes, with and without the use of digital technologies (ACMMG088)

The purpose of this problem is to think about tiling and understand what it is. It also gives students the opportunity to work with basic 2-dimensional shapes and to gain more intuition regarding them, what can be done with them and how the same shapes fit together.

When examining students' examples of tiling, accept any responses where the shapes seem to fit together without gaps. There's no need at this level for students to be able to use angles to justify that the tiles have no gaps between them.

There are clear links between this activity and *Classroom shapes* (p.78) and you may find value in moving from one to the other.

Tessellations are important in architecture and art. You might like to show your students material from the Alhambra in Spain, as well as examples of Escher's work. You can find links on the series website.

Big ideas

» Tiling, including the way shapes fit together

Suggested resources

» Triangular and 4-sided shapes
» Coloured paper and glue

Problem aims
» To see how various 2-dimensional shapes can be used for tiling
» To initiate a concept of infinity

Key concepts
» Using shapes for tiling a floor that goes on forever in all directions
» Tiling requires there to be no gaps between tiles

Possible heuristics/strategies
» Trial and error
» Exhaustive search (try all possible cases)
» Be systematic

Special notes
Tiling: Covering a shape with tiles so that there are no gaps. This is also known as *tessellating*. You may be interested to know that a *tessella* is a small tile used to make a mosaic; in Latin the word means 'small square'. In fact the Latin word comes from the Greek word that means 'four', so presumably the original tiles were mainly squares or rectangles.

Level 1: Hal's floor

Problem

> Horrible Hal's humungous hall
> Is as long and as wide as the sky is tall.
> You can run all day, you can run all night,
> But never the end will come into sight.
> The walls may be red, the walls may be green,
> Nobody knows as they've never been seen.

Can you describe Horrible Hal's humungous hall?

Problem steps

Step 1

This is a pure flight of fantasy. Let students say whatever they want about Horrible Hal's humungous hall, but let them realise that the floor and ceiling go off forever in all directions.

Ask them specifically to tell you what the floor looks like. Have them draw a picture of Horrible Hal standing on his floor.

Step 2

It turns out that the floor in Horrible Hal's humungous hall can be covered in rectangular marble tiles that are all the same size. Can your class say how this is possible? Can they say whether Horrible Hal did this so that there are no gaps anywhere on the floor between tiles?

Let them do this practically by cutting out rectangles and sticking them together on a sheet of paper. How would their work continue forever?

Talk about brick walls and how they are built. Could a brick wall go on forever like Horrible Hal's floor? And without any gaps?

Perhaps one way of showing that it does go on forever with no gaps is to look at Figure 2.9. Horrible Hal could make a strip of bricks that had no internal gaps. Then he could put two strips together without gaps. Then three strips, and so on till he had a brick wall that went on forever in all directions. So he could certainly tile the floor of his hall with rectangular tiles. Hal will use this *strip method* a lot more from here on. By adding together an infinite number of strips of a certain type, Hal can tile floors, ceilings, walls—you name it.

Figure 2.9: A strip of rectangles without gaps

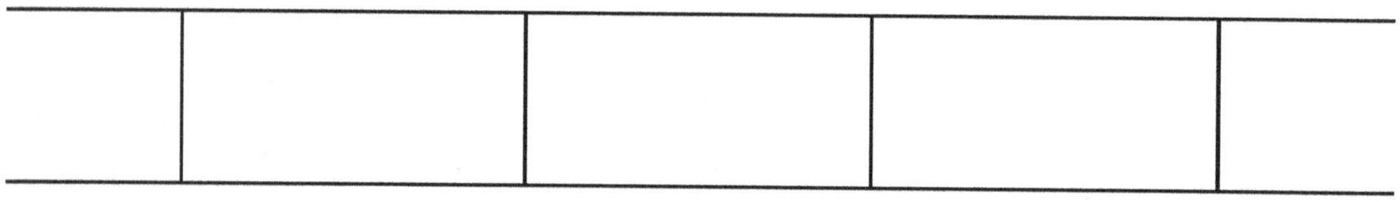

There may be resistance from a few students about whether this is possible; he could never have enough tiles to cover the floor and even if he did, he could never have enough time to put the tiles down. If this does happen, explain that Hal already has a hall that goes on in all directions, so he has no problems about doing anything involving infinite space and time.

Step 3

Ask the class if the hall's floor can be tiled in any other way with rectangular tiles. Let the class work on their own patterns by sticking multiple rectangles together on an A4 sheet.

There are lots of ways to do this (e.g. basket weaves and herringbone), and there are more possibilities if the length of the rectangle is a multiple of the width. As an example, students may get something like the tiling in Figure 2.10.

Figure 2.10: Another way to tile with rectangles

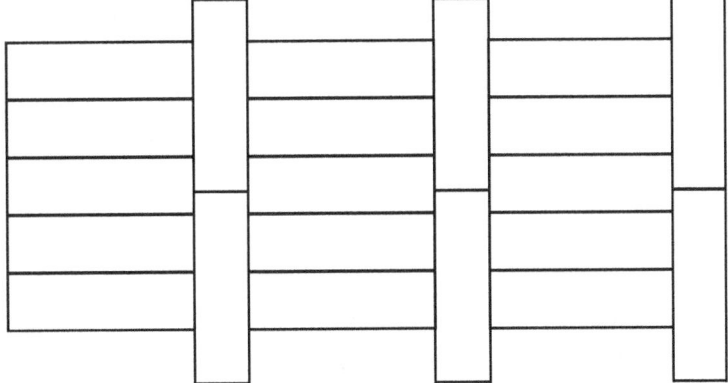

Step 4

Ask the students whether Horrible Hal could tile his floor if he had an infinite number of triangular tiles that were all the same. Suppose that the triangles had three equal sides (equilateral triangles). Could he use these tiles to tile his hall?

Ask them to draw part of such a tiling of the hall. Give a few students a chance to show what they have done and to say why the tiling will have no gaps. The easiest approach is to use the strip method again to demonstrate (see Figure 2.11), as we did with the rectangular tiles in Figure 2.9.

Figure 2.11: A way to tile the hall with triangular tiles

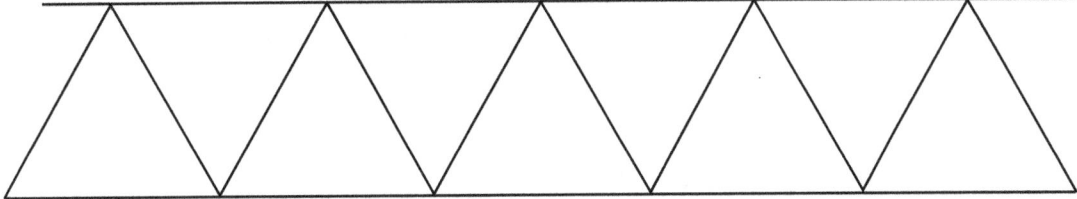

Now give small groups 20 or so triangles that are all the same. Have them stick the triangles onto a piece of A4 paper to show that they fit together and that they could continue in this way until the hall floor was tiled.

Let some students present their work and justify the tilings.

Step 5

Now undertake an investigation. Can Horrible Hal tile the floor in his hall with *any* triangle? (Some possible examples are shown in the *Classroom shapes* activity earlier in this chapter. Use at least four differently shaped triangles here.)

Using the strip approach from above, any triangle can be used to tile the hall's floor (see Figure 2.12). Let the students demonstrate this, either with a drawing or by sticking the same shapes together so that they fit.

Figure 2.12: Tiling with a strange triangle

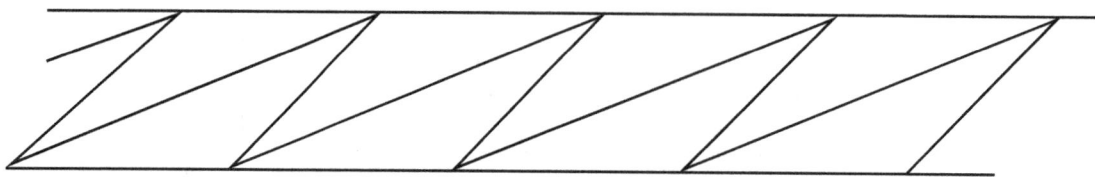

Where to from here?

▌ Could Horrible Hal tile his floor with circles? What other tiles might he try? What tiles definitely won't tile his floor without gaps? What will and why?

▌ What ideas do the students have for tiling floors?

▌ How can they tell when a set of tiles will tile a floor?

Level 2: Hal's ceiling I

Problem

Draw a picture of Horrible Hal looking at his ceiling. Think especially about what the ceiling looks like.

Problem steps

Step 1

Like the floor, the ceiling of the hall is tiled, but with square tiles that are all the same size. Can your class say how this is possible?

Because squares are rectangles we can use the tiling by rectangles from Level 1 to do this. Make sure students can justify the 'no gaps' property of the tiling though. (They may be able to use the fact that four quarter turns add up to a full turn.) Have they seen any real-life examples of tiling using squares?

Step 2

Can your class find some different tilings by squares? Let them experiment. They should be able to find lots based on the rectangular tilings of Level 1.

Step 3

Is it possible to tile Hal's ceiling using tiles that don't have straight edges? (This means the same tile repeated. Let your students experiment here, but they can do this by starting with a square tile and change it a little.) Some examples are shown in Figure 2.13.

Ask your students to tell you the general principle behind this tiling. Make sure that they have checked that the tiles will actually fit without gaps.

Figure 2.13: Tiling with tiles whose edges aren't all straight

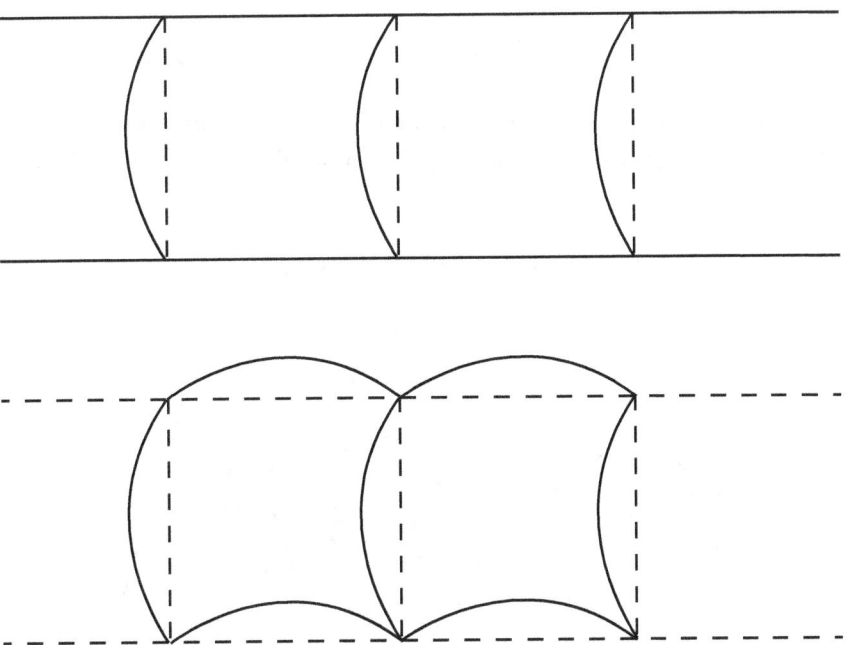

Step 4

Ask your students to find their own tilings with other shapes that don't have straight edges.

Get them to paint these tiles different colours and put them on the walls of your class so the students can compare the shapes.

After they have experimented with these ideas, you may like to introduce them to Escher and his works. From here they could even turn their new tiles into different animals by adding eyes and other details.

Step 5

Can they tile Hal's ceiling with tiles that have five *straight* sides? Let them experiment to see if they can find a way to do this. Will tiling work with *any* 5-sided figure or will it only work for *some*?

It will in fact only work for some 5-sided figures. Figure 2.14 shows a figure that works, along with how it fits together for tiling. Figure 2.15 is an example of something that won't tile the ceiling.

Discuss what they have found and show them the examples below.

Figure 2.14: A pentagonal shape that will tile a flat surface, or *plane*

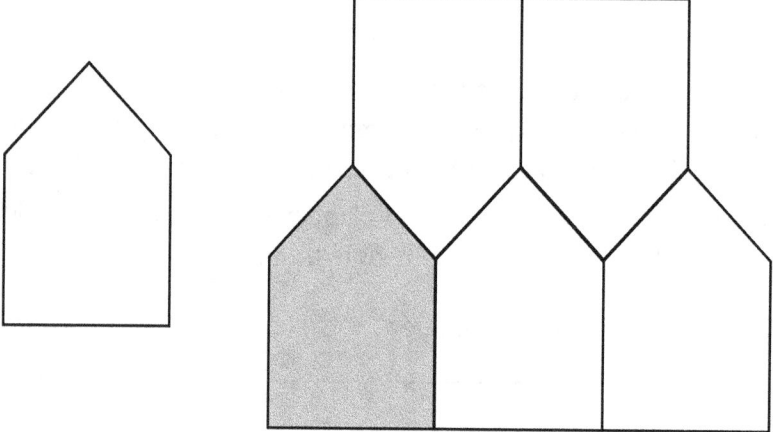

The 5-sided shape in Figure 2.14 *will* tile Hal's ceiling because, as shown, it can be put together to make a strip with parallel sides. This is a bit like the tiles from Step 2 but shifted along a bit.

Figure 2.15: A pentagonal shape that won't tile a plane and why

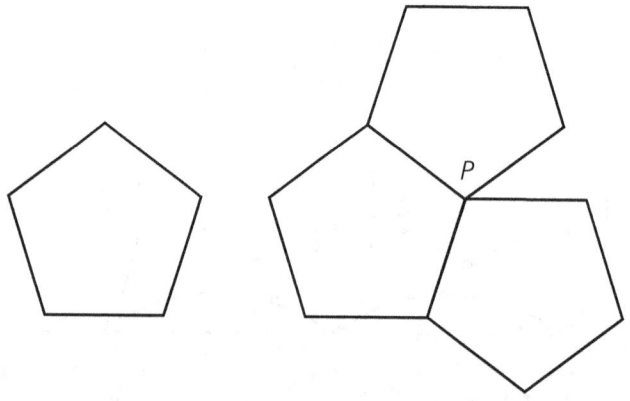

The 5-sided figure in Figure 2.15 *won't* be any good for Hal's ceiling because you can't put three or more together at a corner point *P* so that they fill all of the ceiling around that point.

The crucial thing here is that it must be possible to put other tiles at *all* corners of that shape so that the ceiling is covered in that region.

Step 6

Explain that we call a 5-sided shape with straight sides a *pentagon*. The one in Figure 2.15 has all its angles and sides the same so we call it a *regular pentagon*. In fact, any shape that has all its angles and straight sides the same is called *regular*.

Now explain that a 6-sided shape with straight sides is a *hexagon*. Can your students tile Hal's ceiling with regular hexagons?

Take a vote. If anyone says 'yes' they have to show how it could be done. Those that say 'no' have to justify this.

Figure 2.16: A regular hexagon is made up of six regular triangles

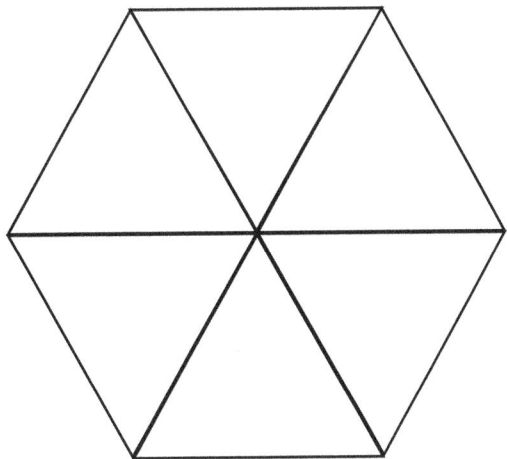

Figure 2.16 shows us that the answer is 'yes' because a regular hexagon is made up of six triangles that are all the same. (In fact they are all equilateral [regular] triangles.)

Ask the class what other hexagonal tiles Hal could use. Could he use *any* hexagon?

Again, some hexagons will tile and some won't. Send the class away in groups to discuss this and report back. They might like to experiment by making shapes for themselves by cutting the shapes out of paper.

Figure 2.17: Some hexagons will tile and some won't

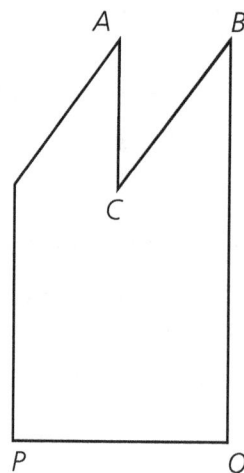

In Figure 2.17, if angles *A*, *B* and *C* are equal and angles *P* and *Q* are quarter turns or right angles, then it will be useful for Hal. On the other hand, if the angle *C* is smaller than both *A* and *B*, then it won't.

What other shapes did the class find?

Where to from here?

- Look at other regularly shaped tiles to see which ones do and which ones don't tile.
- Is there a shape that will tile with seven edges, eight edges, nine edges, ten edges and so on?
- What other questions can students think of that are similar in some way to the questions raised by Hal's ceiling?

Level 3: Hal's ceiling II

Problem

Can Hal use trapezium-shaped tiles to tile his ceiling?

Problem steps

Step 1

In Levels 1 and 2 we saw how to tile Hal's ceiling with triangles, rectangles and squares. Now we look at another quadrilateral: the *trapezium*.

We show one way to prove that trapeziums can be used for tiling in Figure 2.18, repeating the 'strip' approach. Again we are using the strip method here. By putting an infinite number of these strips together, Hal can cover the whole of his ceiling with no gaps.

Figure 2.18: Tiling with a trapezium

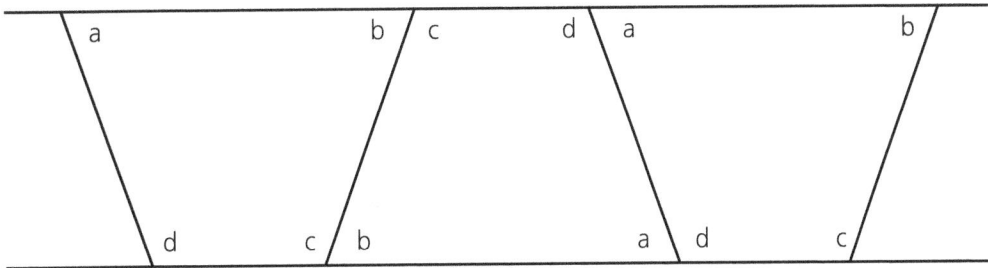

Step 2

Ask your students if Hal can tile his ceiling in more than one way with trapezium-shaped tiles.

This depends on whether you allow Hal to flip the tiles over or not. If he can't, there is only one way; if he can, then there are lots. It all comes down to making sure there are no gaps.

Step 3

Now undertake an investigation of tiling with *any* 4-sided shape. Show the class the picture from *Classroom shapes* (p.78), which you can accessed via the series website. Ask the class to think about whether tilings might be possible or not for each shape. Take a vote on which shapes can tile, and which can't tile, Horrible Hal's ceiling.

Give small groups 20 or so copies of two or more of these shapes and let them try to tile an A4 sheet of paper. Make sure that more than one group has copies of the same shape. Which ones can tile and which can't?

Have a class discussion where students present their results. Whatever answer they come up with for their tiles they should try to justify it in some way. They can all tile using the strip method but some of the strips don't have straight lines. We show a few examples below.

Figure 2.19: Tiling with a parallelogram

Figure 2.20: Tiling with a rhombus

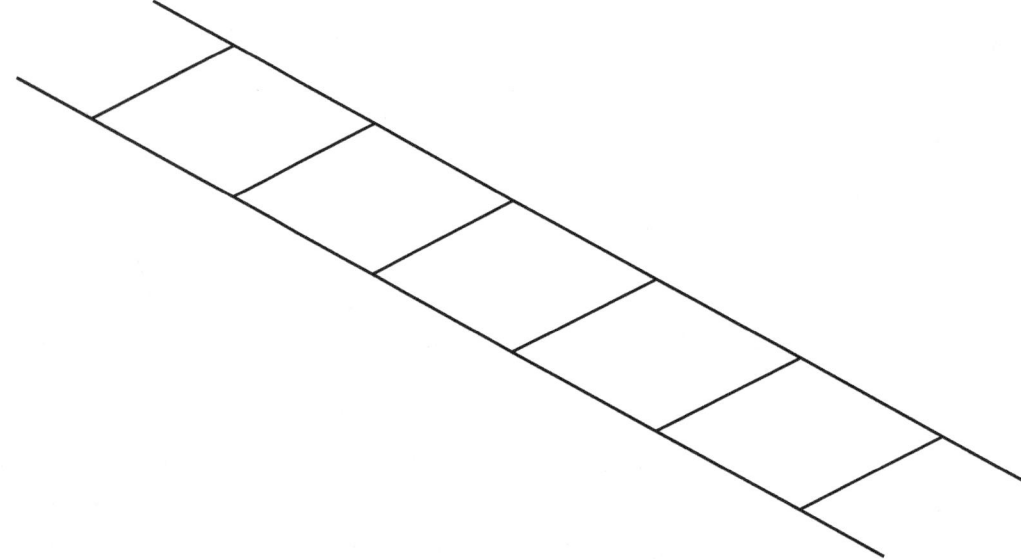

Figure 2.21: Tiling with an unnamed 4-sided shape

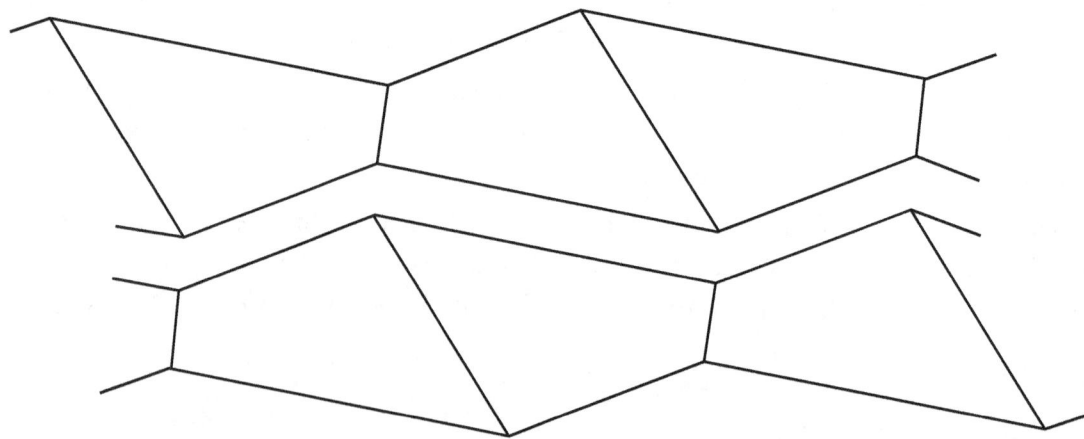

Here we've tried to show how two strips will fit together. We deliberately haven't put them together because the situation is a little more difficult than where the strips have parallel sides (such as with rectangles and rhombuses). Hopefully they can see how these shapes do fit together and so do tile Hal's ceiling.

Figure 2.22 shows the same approach for the dart shape.

Figure 2.22: Tiling with a dart

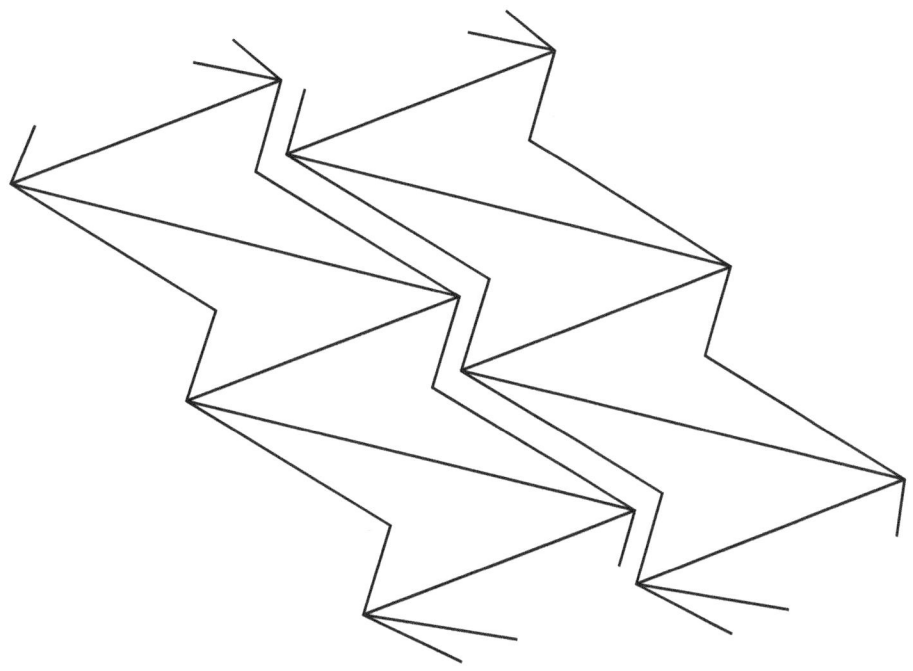

Where to from here?

- Can the students justify the fact that a particular 4-sided shape will tile Hal's ceiling? (It might help to tear the angles off some 4-sided tiles and put them together. Do they make one full turn?)
- Perhaps your students could design a room that has different tessellations on the floor, walls and ceilings.
- What other ideas do your students have?

PART 3: STATISTICS AND PROBABILITY

Part 3 presents three activities based on the Statistics and Probability strand.

Table 3.1: Statistics and Probability activities

Problem	Big ideas
The longest name	• Using pictures (graphs) to display data
	• Interpreting data
Penny's Pet Shop	• How to do some basic counting
The teddies' race	• Beginning to understand chance
	• Thinking about fair games

Some reminders before you use these tasks in your classroom:

1. The questions in the text are ones you can ask your students. You're likely to be able to produce similar, more immediately relevant ones for your particular students as you work on these activities with them.

2. We have given suggested links to the Years in the Australian Curriculum: Mathematics for all the Levels in each activity. Given that there will be a spread of ability in your classes, you should take these as a guide only. Take the opportunity to encourage every student to the edge of their comfort zone.

3. To take all students further, sometimes you can omit some of the later steps of a Level in favour of the early steps in the following Level.

CHAPTER 10:
THE LONGEST NAME

Initial problem

Who has the longest name in the class? Who has the shortest name? Which length of name is the most common? Is there a name of every possible length from 1 to 10?

Ali   Natalie   David   Pippa
            Dominic
Emillia   Elizabeth   Aallyah
     Anthony
              Mackenzie   Fletcher
     Carter
Felix   Lexi   Muhammad   Hazel

Background information

In Level 1 of *The longest name* we look at how long the names of students in your class are. Then we think about how to show or display this so that it is easy to see the differences in length between them, and whose name has a given length. As well as introducing tools for the students to use, we show that the more data that is collected, the more likely we can make a statement about the population of our area. This can be developed across the other classes of that age and then across the whole school. The first few steps of Level 1 are accessible for all students, while Year 2 students should be able to do all of Level 1.

In Level 2 we move on to comparing the lengths of popular names from a few countries. We talk about Australian, English and French names, but you should try to tailor this to your class and choose countries with which some students have an affiliation and for which you can find the richest data. This gives students the opportunity to use the idea of a graph that they have seen in Level 1. The research part of Level 2 will be hard for Foundation and Year 1 students, though if you give them the data they should be able to make some progress. Most students in Year 2 should be able to do this level, as should all students from Year 3 and above.

Level 3 uses the same themes again but this time using the names of students' pets. Level 3 is available to many Year 2 students and all students in Year 3 and Year 4.

Table 3.2: Australian Curriculum content descriptions for *The longest name* activity

Activity level	Problem	Content descriptions
1	Classroom names	*Year 1* Choose simple questions and gather responses (ACMSP262) *Year 2* Identify a question of interest based on one categorical variable. Gather data relevant to the question (ACMSP048)
2	Names here and there	*Year 1* ACMSP262 (see above) *Year 2* ACMSP048 (see above) *Year 3* Identify questions or issues for categorical variables. Identify data sources and plan methods of data collection and recording (ACMSP068) Collect data, organise into categories and create displays using lists, tables, picture graphs and simple column graphs, with and without the use of digital technologies (ACMSP069) Interpret and compare data displays (ACMSP070)
3	Pets' names	*Year 2* ACMSP048 (see above) *Year 3* ACMSP068 (see above) ACMSP069 (see above) ACMSP070 (see above) *Year 4* Select and trial methods for data collection, including survey questions and recording sheets (ACMSP095) Construct suitable data displays, with and without the use of digital technologies, from given or collected data. Include tables, column graphs and picture graphs where one picture can represent many data values (ACMSP096) Evaluate the effectiveness of different displays in illustrating data features including variability (ACMSP097)

The first two levels of this activity aim to explain the basis for all surveys. We provide the reason for this survey (looking at the number of letters in the longest and shortest names and their spread) to let the students know what information to collect. Then we suggest ways of representing their data: using the students themselves, using graphs and using tables. Finally we try to interpret what the data is telling us. These things cover the basic reasons and methods for surveys, which are listed below.

» To answer a question
» To collect appropriate data
» To display the data in a meaningful way
» To interpret the data, then answer the initial question

Note that the thinking and reasoning that goes along with the survey is a fundamental aspect. There is no point collecting data without some reason, and once collected it has to be interpreted.

At this stage the students are unable to use even simple statistical or mathematical techniques. (These will come later.) The conclusions that students may be able to make will therefore be limited, and not just by the data they are able to collect. However, every attempt should be made to make some interpretation of what the data says, even if this is only something like 'there are no names bigger than 10 letters long'.

Big ideas
» Collecting data is to answer questions
» Importance of displaying data

Suggested resources
» Butcher's paper

Problem aims
» To consolidate bigger and smaller
» To consolidate the numbers from 1 to 10
» To introduce the idea of a graph
» To collect data
» To carry out simple surveys
» To interpret data

Key concepts
» Comparing lists in some way

Possible heuristics/strategies
» Collect data
» Act it out
» Draw a picture
» Draw graphs

Level 1: Classroom names

Problem

Who has the longest name in the class? Who has the shortest name? Which length of name is the most common? Is there a name of every possible length from 1 to 10?

Problem steps

Step 1

Discuss the questions and how they might be answered. For instance, how would you find the longest (first) name in the class?

In fact, what does the question mean? Is one name always the same length?

Because we need a foundation to work with, assume that a name is the number of its letters long. (What other approach can you take? Can the children stretch out the letters so that their names all have the same length? This might be an interesting thing to do as an exercise in measurement.)

Can they guess who has the longest or shortest name? Write their guesses on the board.

One way might be to get all the students to write their names on a card. Get them to put the number of letters on the side of the card opposite to their name. They could then arrange themselves in a line so that the shortest names were at one end and the longest at the other.

Because they can't see the whole situation when they are in the line, get them to stick their names in order on a large sheet of butcher's paper. This will lead to something like the image we have shown in Figure 3.1. Tell them that we usually call this type of picture a *graph*.

Figure 3.1: A graph of the names of the children in your class

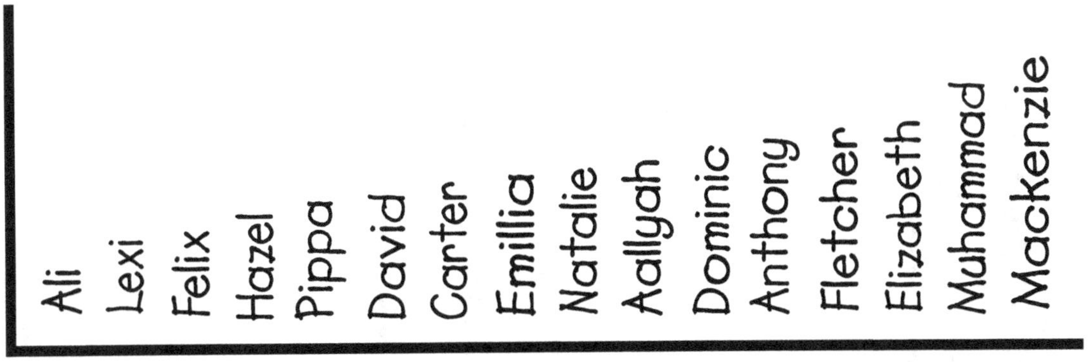

This graph is very useful, because you can tell just from looking at it whose name is the longest, whose is the shortest, and whose name has the most common length. You can also tell if your class has every length from 1 to 10.

Step 2

Discuss why there is no name of length 50.

Also note that some cultures seem to have longer names than others. Why might this be the case?

Step 3

How are Chinese names different from English ones?

If possible, ask a parent with a non-English-speaking background to talk about the way that babies are named in their culture. Does this lead to long or short names?

Step 4

Suppose that there is no name on your graph of length eight. Does that mean that no-one in the world has a name of that length? Can you think of a name that has eight letters?

After thinking about this, what else could you find out from this graph? Are there three names that are the same length? Can you think of three names that are the same?

Is there any difference between the lengths of girls' names and the lengths of boys' names? How would you find this out?

(One way to facilitate this might be for girls' names to be written in green and boys' names in red.)

Step 5

What differences might you expect if you graphed the names of everyone in the same year level as your class?

Discuss how students could do this and what it would be like when they finished it. They'd need to collect name lengths for the other classes (with the cooperation and assistance of those classes' teachers.) Could every child in the year level draw their own name to be added to a big graph? Who should stick the new names on the new, larger graph? Talk about words such as 'data'.

Would it be easier to just put a block on the graph for each name? This way the graph (see Figure 3.2) would be less cumbersome.

Figure 3.2: A histogram with the data from the year level

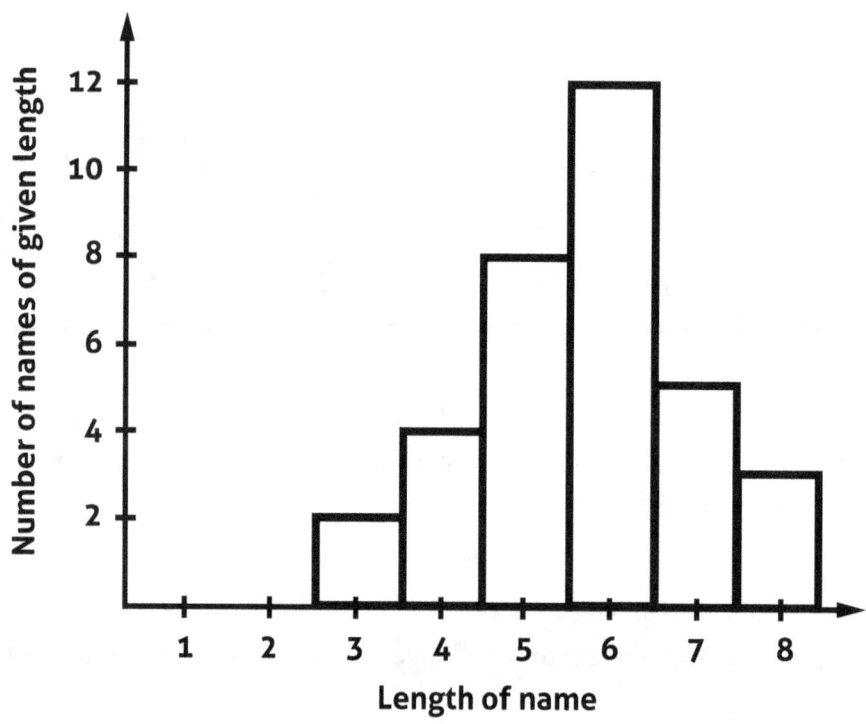

What does the year-level graph tell you that your class graph didn't? Is there now a longer name, a shorter name, or is the name with the most letters a different length? Does this tell you more about the name lengths of children in Australia as a whole?

Note that the data only tells you something about the children in your year level. The longest name that they find for your year level may not be the longest name for all children in the school, the suburb, the state or the country as a whole.

Do the students know how many children are in their year level?

Step 6

Now how about the whole school? Can your students predict the results before they collect the data? What conjectures/guesses can they make?

How could they organise to collect the data? How could they represent the graph so that it was on just one page?

What does this graph tell them that the others didn't? Do we have a better picture of all names in your suburb, your state or the country as a whole?

Are there still some holes in the graph where no names fit? Can they think of a name that would fit in the gap? Can they think of a name that is bigger than the biggest name in the school? Could they make one up?

Where to from here?

- Ask your students what questions started this activity and what answers they came up with.
- Ask them how they showed the data they had collected. Would they use the same methods if they had another question to ask?
- Working in pairs, get students to produce their own graphs on a topic that interests them. They might do a survey of names of the pets in the class. How about the numbers of the houses they have lived in? The height of the children in the class? The number of wins of their favourite sports team? (This could develop over the season.) Minimum and maximum air temperatures over a couple of weeks?
- Let them work on this over a week or more. What do their graphs tell them? Can they explain these things to you?

Level 2: Names here and there

Problem

Is it true that names are longer in England than Australia?

There are only three possibilities:

1. Names in Australia are longer.
2. Names in England are longer.
3. Names in both countries are about the same length.

Which of these is most likely?

Problem steps

Step 1

Discuss with your class what they think is the most likely of the three options. Some students might say that the two countries share the same names so there should be no difference. Others might say that English people's names reflect a longer, more complex history and so their names would be generally longer.

How might your class go about finding the answer to this question? They could try to find a list of all first names used in both countries and compare the lengths, but this would be very difficult because there will be a large number of names involved.

So how can we reduce the data, the number of names we have to look at? There are links on the series website to lists of the most popular 100 names for different countries. This will provide data that the students could graph.

However, even that is a lot of data to handle. To make things more manageable, perhaps they could start with the 10 or 20 most popular names.

Step 2

What result do they find? Is it option 1, 2 or 3?

From reading the links it looks as if, in 2011, the top 20 English names were slightly shorter than the top 20 Australian names. Do they have any comments on this?

Step 3

At this point you could have students go further down the lists for comparison.

Alternatively, have them compare Australian names with American names or even French names, using the links on the series website.

It may come as a surprise that the French names are even shorter than the English names (no hyphenated names, such as Jean-Pierre, are listed in the French top 20 boys' and girls' names).

Step 4

What conclusions can the students draw from their graphs?

Let the children discuss this. Make sure that they understand that they have to be careful about what they say. They can't conclude that Australian names are longer than English or French names. At best they can say that the results are true for the top 20 most popular names in those countries for the years of the data.

Help them to see that because the names were chosen in a special way it narrows the conclusions that they can make.

Where to from here?

- How did your students show the data in the work above? What was the simplest data display for them to understand?
- Popular names can be found online for several countries. Some of your students will have roots in other countries. These students might compare the names of people in that country with Australian names.
- On the other hand, your class might want to try to compare things numerically. In total, the top 20 names that we looked at had 102 letters for French names, 106 letters for English names and 111 letters for Australian names. What can they find out for other countries? How does this numerical way of looking at the data help? (Can your students understand the concept of 'average'?)
- Can your students think of other questions that might be asked about the lengths of names in different countries?

Level 3: Pets' names

Problem

'I've heard some funny names for cats,' Adala said.
'Yes and some unfair ones,' Nusrat replied. 'Fancy calling a cat "Dog" for instance!'
'Really?' asked Adala in amazement.
'Yes! The poor animal must have been scarred for life thinking of having to go on a lead and beg.'
'But a friend of mine called her cat Indigo Blue Joy,' said Adala.
'That seems a bit long to me, certainly bigger than Dog.'
'I wonder how long cat's names usually are,' wondered Adala.

How could they find out how long cats' names usually are?

What conclusions can they make about the lengths of cats' names?

Problem steps

Step 1

Get your class to discuss this conversation that talks about the names of two cats. How could they find out how long cats' names usually are?

This is essentially what was done in Level 1, but now students should be more involved in deciding what to do. Split the class up into smaller groups and ask them to develop a plan to tackle the question. Once each group has a plan, let them go ahead, collect and display their data and draw their conclusions. While this is going on, go round the class and give help and advice as needed.

When this has all been done, get the class together to draw conclusions.

Step 2

Develop the idea further by considering the relative lengths of cats' and dogs' names.

Before the class starts, ask them to conjecture whether cats' names are longer, shorter or equal in length to dogs' names.

Repeat the elements of Step 1.

Step 3

Repeat Step 2 using cats', dogs' and humans' names.

What conclusions did they reach? Can they explain these conclusions?

Where to from here?

- What was the most interesting conclusion that they reached?
- What method of showing their data was least useful?
- Can they think of other questions like the ones here?

CHAPTER 11: PENNY'S PET SHOP

Initial problem

Penny's Pet Shop has two large glass boxes to display the live animals it sells. One of the boxes is for kittens and one for puppies. At the moment Penny has 10 animals altogether.

Can the class draw a picture of Penny's glass boxes with the animals in them?

Background information

The aim of *Penny's Pet Shop* is to introduce the idea of *systematic counting* as a basis for theoretical probability later. This activity has links to the Number and Algebra strand.

Level 1 looks at how many ways 10 kittens and puppies can be shown in the pet shop's two glass display cases. (This also considers how you can make 10 by adding two numbers together.) You could use a variant of this problem to introduce any new number to your class. All students can do at least the early parts of Level 1, especially if the number of animals is reduced. We expect the whole level to be achievable by Year 2 students and above.

»

Level 2 repeats the outline of Level 1 but looks at another type of grouping. In doing so, we extend the problem to three types of animals. This counting is more difficult than before, but it can be simplified by referring back to what was done in Level 1. Level 2 can be started by Year 1 students, but this is really an activity for Year 2 students and above.

Level 3 extends the idea of Level 2 by looking at various numbers of animals in three boxes and trying to find a pattern. At all levels we look at the chance of some number of animals being in Penny's shop. Level 3 can be started by Year 2 students, but the full activity is designed for Year 3 and Year 4 students.

Table 3.3: Australian Curriculum content descriptions content for *Penny's Pet Shop*

Activity level	Problem	Content descriptions
1	Ten animals	*Foundation* Compare, order and make correspondences between collections, initially to 20, and explain reasoning (ACMNA289) *Year 1* Identify outcomes of familiar events involving chance and describe them using everyday language such as 'will happen', 'won't happen' or 'might happen' (ACMSP024) Investigate and describe number patterns formed by skip counting and patterns with objects (ACMNA018) *Year 2* Identify practical activities and everyday events that involve chance. Describe outcomes as 'likely' or 'unlikely' and identify some events as 'certain' or 'impossible' (ACMSP047) Describe patterns with numbers and identify missing elements (ACMNA035)
2	Three display boxes	*Year 1* ACMSP024 (see above) ACMNA018 (see above) *Year 2* ACMSP047 (see above) ACMNA035 (see above) *Year 3* Conduct chance experiments, identify and describe possible outcomes and recognise variation in results (ACMSP067) Describe, continue, and create number patterns resulting from performing addition or subtraction (ACMNA060)
3	More than six animals	*Year 2* ACMSP047 (see above) ACMNA035 (see above) *Year 3* ACMSP067 (see above) ACMNA060 (see above) *Year 4* Describe possible everyday events and order their chances of occurring (ACMSP092)

It will take some time to do all of the problems and extensions here, perhaps even most of the year on and off. Don't rush any of it; give students time to internalise and become confident in the concepts before exploring further.

This activity looks like a number exercise, especially at Level 1, but it's also about the basis of theoretical probability. Even Level 1 looks at descriptions of chance in terms of 'will happen', 'won't happen' and so on. It also provides the beginning of ideas that will consolidate into concepts such as outcomes and event space.

This problem has obvious connections to *The farmyard problem* activity of Chapter 3 (p.56).

Big ideas

Addition of numbers to 10; basic ideas of chance/probability: some things are more likely to happen than others; why some things are more likely to happen than others.

Suggested resources

- Paint or pens
- Paper
- Stamps of puppies or kittens
- Pictures of puppies and kittens cut out of books

Problem aims

- To develop the concept of 10
- To ensure that the whole class can count up to 10
- To start to develop the basic number facts relating to 10
- To introduce the number 0
- To begin to see ways to find the sizes of specific groups
- To discuss activities that involve chance

Key concepts

- Being systematic
- Describing outcomes as 'likely, 'unlikely' and so on

Possible heuristics/strategies

- Use something that you've met before
- Make a drawing
- Make a table
- Be systematic

Level 1: Ten animals

Problem

Penny's Pet Shop has two large glass boxes to display the live animals it sells. One of the boxes is for kittens and one for puppies. At the moment Penny has 10 animals altogether. Can the class draw a picture of Penny's glass boxes with the animals in them?

Problem steps

Step 1

Have the children draw pictures of the glass boxes and then draw the animals inside them.

The specific numbers of kittens and puppies have been deliberately omitted from the problem. The children can produce any picture that has 10 animals; Figure 3.3 shows one example.

Figure 3.3: Four kittens and six puppies make up 10 animals

Allow the children time to check that they have 10 animals.

Choose a child to show their picture. Get everyone to make sure that the picture is an answer to the problem.

Ask 'Did anyone else draw a picture of four kittens and six puppies?' Let all the children who have four kittens and six puppies come to the front with their drawings. Check that the drawings do in fact have four kittens and six puppies. Put all of these pictures together on a board or wall with 'four kittens and six puppies' written underneath. (Or even 4 + 6 = 10, if your students are ready for that.)

What did the other children draw? Did any of them draw four puppies and six kittens? Put all those pictures on the board/wall with 'four puppies and six kittens' written underneath.

Continue in this vein until everyone's drawing is on the wall. Make sure that the pictures are placed systematically, starting with the fewest number of kittens and working up to the biggest number.

Step 2

Are there any possibilities that no-one has drawn? Discuss this with the children. What drawings are needed to produce *all possible* answers to the original problem?

Lead them, if necessary, to the case of 'no kittens'. Talk about the number zero. Make sure that the zero puppies case is also included. (If you think that your children are not ready for zero yet, then avoid it at this stage.)

Ask them if there are any more possibilities? How many ways are there to draw the kittens and puppies in Penny's Pet Shop?

Continue until all of the possibilities have been listed: 0 + 10; 1 + 9; 2 + 8; 3 + 7; 4 + 6; 5 + 5; 6 + 4; 7 + 3; 8 + 2; 9 + 1; and 10 + 0. There should be 11 possibilities (or nine if you plan to avoid zero).

Step 3

Discuss the link between the animals and the number of ways that numbers can add to 10. There is one more possibility than there are animals. Why? (Because there are 11 combinations of kittens and puppies.)

Discuss their explanations.

(Note that if you're avoiding zero, there are nine combinations of kittens and puppies, or one fewer possibility than there are animals, so you will need to adjust discussions accordingly.)

Step 4

Ask some simple probability questions such as those listed below. To find the answers refer to the students' drawings.

- What are the chances of finding 10 animals in Penny's Pet Shop? (Will happen; it's certain that you will.)
- What is the chance of finding 11 animals in the two glass boxes for puppies and kittens? (Won't happen; no chance or impossible.)
- What is the chance of finding more than five kittens or more than five puppies? (Might happen; likely.)
- What is the chance of finding *no* kittens? (Might happen; very unlikely.)
- Is it more likely for there to be *more* than two puppies or *less* than two puppies? (More than two puppies is more likely.)

Step 5

Get each student to make a list of things that:

- *will* happen; it's certain that they will
- *might* happen; they are *likely* though
- are *more likely* than something else
- *might* happen; they are *unlikely* though
- *won't* happen; there is no chance, they are impossible.

Give some children the opportunity to ask the class these things. Let each child give their answer and then the class can discuss what they have said.

Put students into groups and have them make up their own set of 10 probability questions. They could then challenge other groups to a quiz.

Step 6

Change the parameters of the initial problem from 10 animals to any other number of animals that you like. Repeat some of the questions and activities above. These should include getting all possible contents of Penny's glass boxes and some thinking about probability.

Step 7

Suppose that Penny has 15 animals. Can your students predict how many ways Penny might have this many kittens and puppies in her two boxes?

Suppose that you have any specific, large number of animals (say 200—they're *big* glass boxes). Can the students predict the different numbers of ways there are of putting that many kittens and puppies into the boxes? (One more than the number of animals if you let zero be used; one less if you don't.)

If they can correctly conjecture the answer to this for any number in their range of numbers, then they have reached a generalisation.

Where to from here?

- How did the students work out how many ways Penny can arrange 15 animals in the two boxes?
- Try changing the context while keeping the problem essentially the same. For example, the Jung family has four children. How many ways can this happen in terms of the number of boys and girls they might have?
- Follow up this problem by getting the class to think about what might happen if there were more boxes for the kittens and puppies.
- You might even ask the class to think about what happens if there are more than two kinds of animals (see Level 2).

Level 2: Three display boxes

Problem

What happens if Penny's Pet Shop has three glass display boxes for kittens, puppies and rabbits? Can they draw a picture of a day when Penny has five animals on display? How many possible ways are there for Penny to display five animals?

Problem outline

Step 1

Discuss how your class may tackle this problem.

You could use the same approach as the Level 1 problem; however, there are 21 different possibilities, so it's better to be systematic. It may also be worth using a schematic drawing to save time; for instance, using K for kitten, P for puppy and R for rabbit to show the various combinations. Figure 3.4 demonstrates one way of doing this. Here we look at the possibilities for 0, 1, 2, 3, 4, and 5 rabbits.

You could start by talking students through the possibilities for no rabbits. Then get them to work systematically through the other possible cases on their own.

Figure 3.4: Combinations of kittens, puppies and rabbits to make five animals altogether

No rabbits:	CCCCC, CCCCD, CCCDD, CCDDD, CDDDD, DDDDD	6
One rabbit:	CCCCR, CCCDR, CCDDR, CDDDR, DDDDR	5
Two rabbits:	CCCRR, CCDRR, CDDRR, DDDRR	4
Three rabbits:	CCRRR, CDRRR, DDRRR	3
Four rabbits:	CRRRR, DRRRR	2
Five rabbits:	RRRRR	1
	Total	**21**

It is important to be systematic in taking the number of rabbits in order. Being systematic ensures that no possibility is missed.

Step 2

Can anyone see another way to do this problem? Do the numbers 6, 5, 4, 3, 2, 1 remind the class of anything?

Refer back to the Level 1 activity's findings to explain that:

- 6 is the number of ways that 5 animals can be put into two glass boxes
- 5 is the number of ways that 4 animals can be put into two glass boxes
- 4 is the number of ways that 3 animals can be put into two glass boxes
- 3 is the number of ways that 2 animals can be put into two glass boxes
- 2 is the number of ways that 1 animals can be put into two glass boxes
- 1 is the number of ways that 0 animals can be put into two glass boxes.

So all you have to do is to add 6 to 5 to 4 to 3 to 2 to 1 to get the answer for five animals in three boxes.

Step 3

Explore the likelihood of certain combinations/discoveries happening. Use a variety of questions such as these:

- What are the chances of finding five animals in Penny's Pet Shop? (Will happen; it's certain that you will.)
- What is the chance of finding six animals in the two glass boxes for kittens and rabbits? (Won't happen; no chance or impossible.)
- What is the chance of finding more than three kittens or more than three puppies? (Might happen; likely.)
- What is the chance of finding no kittens?(Might happen; unlikely.)
- Is it more likely for there to be *more* than two puppies or *less* than two puppies? (More than two puppies is more likely.)

Step 4

Get each student to make a list of things that:

- *will* happen; it's certain that they will
- *might* happen; they are *likely* though
- are *more likely* than something else
- *might* happen; they are *unlikely* though
- *won't* happen; there is no chance, they are impossible.

Give some children the opportunity to ask the class about one item from their list. Let each child give their answer and then the class can discuss what they have said.

Where to from here?

- What did the students find was the hardest problem here and why?
- Where did the class use a systematic way of solving this problem?
- Discuss what questions could be asked next.

Level 3: More than six animals

Problem

Suppose that Penny's Pet Shop has three large display boxes for kittens, puppies and rabbits. Can your students draw a picture of a day when Penny has six animals?

Problem steps

Step 1

How many possible ways are there for Penny to display six animals?

If your students have not done Level 2 they may need some guidance in getting started. If they know how to do the problem, let them start straight into it (with possibly some scaffolding to remind them what they've done before). Let them use any method they're happy with to solve the problem, though the systematic way is best. (The answer is shown in the next step.)

Step 2

How many possible ways are there for Penny to display six animals? What about 7 or 8 animals?

The answers are shown below. The numbers in the additions are produced using the methods of Step 1 in Level 2.

- With 6 animals we have 7 + 6 + 5 + 4 + 3 + 2 + 1 = 28 possibilities.
- With 7 animals we have 8 + 7 + 6 + 5 + 4 + 3 + 2 + 1 = 36 possibilities.
- With 8 animals we have 9 + 8 + 7 + 6 + 5 + 4 + 3 + 2 + 1 = 45 possibilities.

Step 3

Ask the class if they can see any patterns in the numbers of the last step.

They will probably notice that the numbers are going up in steps. In the list above the total increases by 8 in going from 6 to 7 animals, and increases by 9 in going from 7 to 8 animals.

If there were 9 animals, would the increase be 10? Can they think why? (It's the number you get if there are no rabbits. The other numbers: 9 you get with 1 rabbit and 8 of the others; 8 you get with 2 rabbits and 7 of the others; 7 you get with 3 rabbits and 6 of the others; and so on.)

You might also show them that this can be done by multiplying the number of animals by the number of animals plus one, and then dividing by 2. For instance, to find the total possibilities for 9 animals we multiply 9 by 10 and divide by 2. This is harder to explain.

Step 4

Get each student to again make a list of things of different likelihood; things that:

- *will* happen; it's certain that they will
- *might* happen; they are likely though
- are more likely than something else
- *might* happen; they are unlikely though
- *won't* happen; there is no chance, they are impossible.

Discuss this in class. Let them together find some things that might happen that would fit into each one of these categories.

Step 5

Put students into groups to make up their own questions about the chances of things happening. Start with the six animal situation and let students make up their own questions for seven and eight animals. They might come up with something like the following:

- What are the chances of finding six animals in Penny's Pet Shop? (Will happen; it's certain that you will.)
- What is the chance of finding seven animals in the two glass boxes for kittens and rabbits? (Won't happen; no chance or impossible.)
- What is the chance of finding more than four kittens or more than four puppies or more than four rabbits? (Might happen; likely.)
- What is the chance of finding no kittens? (Might happen; unlikely.)
- Is it more likely for there to be *more* than three puppies or *less* than three puppies? (More likely than some other things.)

Give some children the opportunity to ask the class one thing from their list. The students asking the questions should be able to justify their answers.

Where to from here?

- What questions here does the class think produced the most discussion? Why?
- What did the students find was the hardest problem here and why?
- The students might look at four types of animals. Can your class work out the possible combinations from knowing about three types of animals?
- Discuss what questions they could ask next.

CHAPTER 12:
THE TEDDIES' RACE

Initial problem

Two teddies are racing along the track. Someone tosses a coin. If it comes up heads, then the girl teddy moves on one space. If it comes up tails, then the boy teddy moves on one space. Who will win the race?

Table 3.4: The teddies' racing track

Girl teddy (heads)	1	2	3	4	5	**Finish**
Boy teddy (tails)	1	2	3	4	5	**Finish**

Background information

The aim of this activity is to provide students an opportunity to develop their intuition regarding chance events. We do this by looking at several race games, some of which are biased and some fair. At no stage do we quantify any probability; we're trying to give students an implicit feel for the probability of events and for fairness.

We start off with a simple race with two teddies in Level 1. If a coin lands heads a particular teddy advances; if it lands tails, the other moves on. Who will win? This is repeated with three teddies using stick dice. What outcomes are likely? Foundation and Year 1 students can play a game like this even if they can't toss a coin properly, so they should be able to do at least the first two steps of the Level 1 activity. These students should also be able to cope with the stick dice in Steps 5 and 6, and in fact will probably enjoy decorating these dice. It's worth discussing fairness with them, though not at any deep level. Year 2 students and above should be able to do the whole level.

These questions are repeated in Level 2 using a die and different numbers of teddies. Year 1 students will be able to play the game but overall it was designed for Levels 2 and above.

The race and chance ideas in Level 3 become harder by using 11 teddies and two dice. Because of the complications of using two dice, the Level 3 activity is most likely more achievable for Years 3 and 4.

In both Levels 2 and 3 the aim of the game is to help students see which games are fair and why, and to develop their ideas of chance.

Table 3.5: Australian Curriculum content descriptions for *The teddies' race*

Activity level	Problem	Content descriptions
1	Coins and sticks	*Year 1* Choose simple questions and gather responses (ACMSP262) *Year 2* Identify a question of interest based on one categorical variable. Gather data relevant to the question (ACMSP048) Collect, check and classify data (ACMSP049)
2	One die	*Year 1* ACMSP262 (see above) *Year 2* ACMSP048 (see above) ACMSP049 (see above) *Year 3* Conduct chance experiments, identify and describe possible outcomes and recognise variation in results (ACMSP067) Collect data, organise into categories and create displays using lists, tables, picture graphs and simple column graphs, with and without the use of digital technologies (ACMSP069) Interpret and compare data displays (ACMSP070)
3	Two dice	*Year 2* ACMSP048 (see above) ACMSP049 (see above) *Year 3* ACMSP067 (see above) ACMSP069 (see above) ACMSP070 (see above) *Year 4* Describe possible everyday events and order their chances of occurring (ACMSP092)

The teddies can be replaced with a variety of toys or even by children to suit your situation.

There are links to 'dice rollers' and card shufflers on the series website that will allow you to construct the equivalent of spinners for various sizes. These sites might be handy and more convenient than actual coins, dice or spinners.

Big ideas

» Basic ideas of chance/probability

» Not every game or situation is fair to all players

Suggested resources

» Craft sticks

» Dice

» Coins

» Spinners

» Teddies or toys to move

Problem aims

» To see how chance works in playing a game

» To see what events are more likely than others

Key concepts

» To see what events have the same chance of happening

» To judge what games are fair games

Possible heuristics /strategies

» Make a table

» Collect data

Special note

A *fair game* gives each player an equal chance of winning.

Level 1: Coins and sticks

Problem

Two teddies are racing along the track. Someone tosses a coin. If it comes up heads, then the girl teddy moves on one space. If it comes up tails, then the boy teddy moves on one space. Who will win the race?

Table 3.6: The teddies' racing track

Girl teddy (heads)	1	2	3	4	5	Finish
Boy teddy (tails)	1	2	3	4	5	Finish

Problem steps

Step 1

The student activity handout on the series website has a copy of the race track.

Run the race in front of the class so that everyone can see how it works. Use two of the children as teddies if you don't have two teddy toys. Get a third child to flip the coin. Which teddy won?

Step 2

If the race is run again, will the same teddy win?

Take a vote: those who think the same teddy will win versus those who think the other teddy will win.

Run the race a few more times, then see if anyone wants to change their vote.

Step 3

Give groups of 3–4 students a game board, a coin and a result sheet (Table 3.7). Let them run the race twelve times, keeping a record of which teddy wins each race.

Table 3.7: Recording results of the races

Race	1	2	3	4	5	6	7	8	9	10	11	12	Teddy 1 wins	Teddy 2 wins
Winner														

Step 4

Who do they think has the best chance of winning, the girl teddy or the boy teddy?

Ask them to justify their choice. It is likely that different groups will say different teddies. Some groups may say that it's about equal.

Combine every group's results. This should show that there's very little difference between the number of times the girl teddy wins and the number of times the boy teddy wins.

If this is the case, it looks as if this is a *fair game*.

Step 5

Take a break to make *stick dice*.

Give each student two flat craft sticks, as well as crayons or paint to decorate them with. They can be decorated however the students want, but *only on one side*.

When the sticks are completed, let the children throw them in the air (toss them). Discuss what the children see. They should notice that sticks come up in four ways: the two decorated sides up; one decorated side up; the other decorated side up; and no decorated sides up.

Return to the race game, this time with three teddies racing. The first teddy moves when the two decorated sticks are up; the second teddy when only one decorated side is up (it doesn't matter which side); and the third teddy when no decorated sides are up. Run a couple of races at the front of the class so that everyone understands the rules.

Step 6

Get every group to race the teddies 12 times. Collect the data (see Table 3.8) to see if this game is fair.

Table 3.8: Racing with stick dice

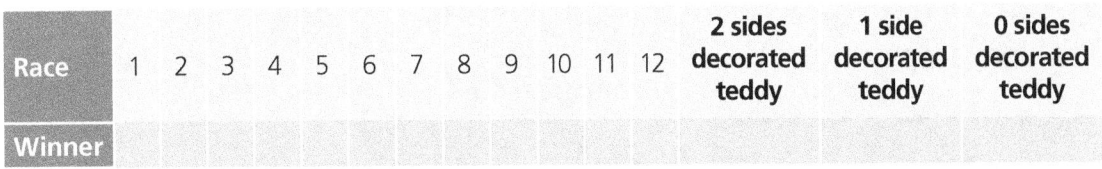

Does the game seem to be fair? Ask each group which teddy they would prefer to be and why. Is that the same for each group? Is that the same when the whole class adds their data together?

Discuss this situation. Could they change the stick race game to make it fair? How would they make it fair? This can be done by adding another teddy and letting the four teddies move this way:

- One teddy moves if 2 decorated sides come up.
- One teddy moves if decorated side A comes up.
- One teddy moves if decorated side B comes up.
- One teddy moves if 0 decorated sides come up.

Why would this game be fair? Talk about what ways the sticks could land. Would any of these ways land more often than any other? Why?

Where to from here?

- What games do students play at home that are *not* fair? Why do they think they are not fair? How could students make the games fair?
- What other race games can the class invent?
- Is it easier to invent fair games or non-fair games?

Level 2: Dice

Problem

Six teddies are racing along the track in Table 3.9. Someone rolls a six-sided die. If a 1 comes up, then Teddy 1 moves on one space. If a 2 comes up, then Teddy 2 moves on one space. The same thing happens for 3, 4, 5 and 6. The teddies only moves forward one square if their number comes up on the die.

Who will reach **Finish** first? Why?

Table 3.9: Six teddies and a die

Teddy 1	1	2	3	4	5	Finish
Teddy 2	1	2	3	4	5	Finish
Teddy 3	1	2	3	4	5	Finish
Teddy 4	1	2	3	4	5	Finish
Teddy 5	1	2	3	4	5	Finish
Teddy 6	1	2	3	4	5	Finish

Problem steps

Step 1

Divide the class up into small groups. Let them play the game several times and record the winners (see Table 3.10). Is this game fair?

Table 3.10: Results for the 6-teddy race

	Teddy 1 wins	Teddy 2 wins	Teddy 3 wins	Teddy 4 wins	Teddy 5 wins	Teddy 6 wins
Number of times						

Let them play the game several times and ask them 'Who will reach **Finish** first? Why?'

Discuss their conclusions.

Step 2

Play the original two-teddy game again, but with a die. If the die shows 2 or less, the girl teddy moves; if the die shows 3 or more the boy teddy moves.

Play the game several times. Who is most likely to win and why?

Step 3

Play a three-teddy game from Level 1, using a die, but let the students decide when each teddy moves forward.

Can they see when a game is going to be fair and when it isn't? Collect data to see if their conjectures are true or not.

Step 4

Repeat Step 3 with a four-teddy game.

Which teddy is most likely to win and why? Is the game they have invented a fair game? Play the game to see.

Step 5

Repeat the last two steps using stick dice.

Where to from here?

- How do your students know when a game is going to be fair or not?
- What do they think makes it fair?
- Let your students invent their own games. Can they predict which ones are fair or not?

Level 3: Two dice

Problem

Eleven teddies are racing along the track of Table 3.11.

Someone rolls two dice. If the total is 2, Teddy 2 moves on one space. If the total is 3, Teddy 3 moves on one space. The same thing happens for 4, 5, 6 up to 12. Each teddy only moves forward one square if their number comes up on the dice.

Who will win the race?

Is the game fair or not?

Table 3.11: An 11-teddy track

Teddy 2	1	2	3	4	5	Finish
Teddy 3	1	2	3	4	5	Finish
Teddy 4	1	2	3	4	5	Finish
Teddy 5	1	2	3	4	5	Finish
Teddy 6	1	2	3	4	5	Finish
Teddy 7	1	2	3	4	5	Finish
Teddy 8	1	2	3	4	5	Finish
Teddy 9	1	2	3	4	5	Finish
Teddy 10	1	2	3	4	5	Finish
Teddy 11	1	2	3	4	5	Finish
Teddy 12	1	2	3	4	5	Finish

Problem steps

Step 1

Play the game to see who wins and to make sure that everyone understands the rules.

Play the game again. Did the same teddy win? Who will win next time?

Take a vote: is this a fair game?

Step 2

Let groups of students work on the game and record their results.

When they have all finished, see if they now think it is fair or not. Discuss the reasons for their answers.

Step 3

If the game is not fair (it isn't), which teddy do they think will have the best chance of winning? Why?

(Teddy 7 will win more often because 7 comes up most often when two dice are rolled.)

Step 4
Now play three variations of the two-teddy game using two dice.
- Variation 1: Teddy 1 moves forward if the total is even, otherwise Teddy 2 moves.
- Variation 2: Teddy 1 moves forward if the total is a multiple of three, otherwise Teddy 2 moves.
- Variation 3: Teddy 1 moves forward if the total is a prime number, otherwise Teddy 2 moves.

Before the students play, get them to say whether they think that the game is fair or not. Discuss each game after the students have played them.

Step 5
For all of the games at this level, ask them to say what things fit the following categories.
- *Will* happen
- *Might* happen
- Are *more likely* to happen than some other things
- *Aren't likely* to happen
- *Won't* happen

Where to from here?
- Would they rather play a game that was fair or unfair?
- Let the class invent their own games using coins, dice, spinners or anything else to see if the games are fair or not.
- This game could be played with a different number of teddies and using stick dice with more than two sticks.

CREATIVE ACTIVITIES IN **MATHEMATICS**

Problem-based learning is a powerful alternative to drill-and-practice or skills-based learning, especially within mathematics.

The *Creative Activities in Mathematics* series provides a wealth of investigations and open-ended active learning activities, designed to engage students with mathematics and develop their problem-solving, collaboration and mathematical skills.

The three titles in the series provide a variety of class activities suitable for students from lower primary to middle secondary, along with teaching notes and staged lesson plans. Each activity is a whole-class investigation with open-ended answers that takes a particular scenario and develops it over multiple levels. This enables it to be used both at different year levels and with students of differing ability in the same class. All activities are firmly grounded in the Australian Curriculum: Mathematics.

Links to extra information, activities and student worksheets are available and easy to access online.

About the authors

Derek Holton is a mathematician and an Honorary Professor at the Melbourne Graduate School of Education.

Cath Pearn is a Senior Research Fellow in the ACER Institute and a lecturer in Mathematics Education at the University of Melbourne.

Duncan Symons is a Lecturer of Science and Mathematics Education at the University of Melbourne.

Charles Lovitt has directed several Australian national and state mathematics projects and is now a consultant and workshop presenter.

Amba Press | www.ambapress.com.au | hello@ambapress.com.au

www.ingramcontent.com/pod-product-compliance
Lightning Source LLC
Chambersburg PA
CBHW081102070526
44584CB00021B/3176